Essential Trigonometry

Made Simple for Middle & High Schools

All essential concepts of trigonometry for the middle and high school students are covered herein.

All core concepts of trigonometry are consolidated together with examples and exercise for effective learning. Concepts explained in clear and simple way together with examples act like self-teaching guide. Each section is included with exercises and solutions in workbook style to ensure guaranteed learning.

Guaranteed learning of trigonometry for the students of middle & high schools

Recommended Age Group – 11 to 17 Years

- **Saurya Singh**

Kalisey Academy Publication

Kalisey Series

Delivering the highest quality academic materials

About the Author

Saurya Singh was born on 17.Nov.2004 in Frankfurt am Main. Right from his childhood, he enjoyed being involved in math-puzzles, Mathletics, Abacus etc. and succeeded in several international math competitions. He had great passion for mathematics and physics performing very well in advanced level of high school and college level at a quite early age.

Saurya also conducts workshops of math and physics and enjoys being involved in blogs. In his books, blogs and workshop, he encourages the enquiry based knowledge and promotes the transdisciplinary approach of learning. His approach stimulates the interest and nurtures the skills in physics and math.

Among extra-curricular activities, Saurya is very much active in sports and music. He is a fantastic badminton player. His other activities include swimming, martial-arts and biking. Saurya loves violin and has been playing since his childhood.

Colophon

Title: Essential Trigonometry - Made Simple for Middle and High Schools

Description: The book presents all core concepts of trigonometry. All core concepts are explained in clear and simple way together with examples. Each section is included with exercises and solutions in workbook style to ensure guaranteed learning.

Author: Saurya Singh

Series: Kalisey Series

Publisher: Kalisey Academy Publication

Book Presenter: Author & publisher

Copyright: Author & Publisher

Website: www.kalisey–softek.com/trigonometry.html

Contact: info@kalisey-softek.com

07.June.2016 - First release; 24.Feb.2017 - First updated version;

**Dedicated to my parents, teachers & my school –
Metropolitan School Frankfurt**

For any further enquiries about the book and content, contact:
info@kalisey-softek.com

Contents

Preface

I learnt trigonometry at a quite early age because it was made simple and interesting to me. Realizing the vastness and complexities of trigonometry, I consider myself lucky that I learnt it with great fun. While working on trigonometry, I always felt like I am solving puzzles instead of complex functions.

I wish to present trigonometry to other students of middle and high schools in such a way that they will find it easy and interesting to learn the concept.

All essential concepts of trigonometry for middle and high school students are covered in this book.

All core concepts of trigonometry are presented here as crash course for efficient learning. Concepts explained in clear and simple way together with examples act like self-teaching guide. Each section is included with exercises and solutions in workbook style to ensure guaranteed learning.

I am sure that students will have guaranteed learning of all core concepts of trigonometry. Readers will find it easy to learn and interesting to move forward, take tougher challenges and start playing with trigonometry.

Have fun learning trigonometry!

- *Saurya Singh* (Frankfurt am Main, *3.June.2016*)

1 Chapter 1: Triangle Overview, Propositions & Theorems

1.1 Triangle

We all know about it but it's better to recall it once more. A triangle is a polygon with three edges and three vertices. The angles of a triangle always add up to 180°; So simple.

1.1.1   Shape & Size of a Triangle

A triangle is a polygon with three edges and three vertices. A triangle with vertices *A*, *B*, and *C* is denoted as $\triangle ABC$

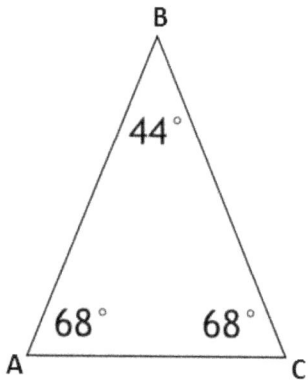

Triangle

The sum of the three angles of a triangle always add up to 180°

<u>Let's recall some of triangles with respect to their sides:</u>

Equilateral triangle: All sides are equal, all angles are 60 degrees

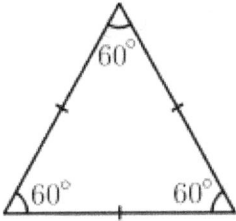

Equilateral triangle

Isosceles triangle: Two sides are equal and two base angles are equal

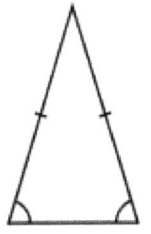

Isosceles triangle

Scalene triangle: All sides are of different length

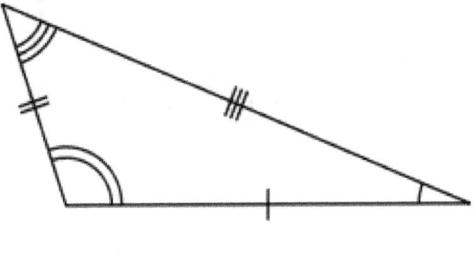

Scalene triangle

<u>Let's recall some of triangles with respect to their internal angles:</u>

Right triangle: One angle is 90 degrees

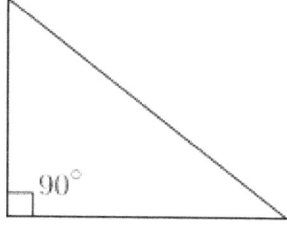

Right triangle

Obtuse triangle: One angle is greater than 90 degrees

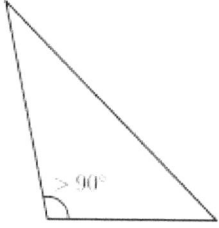

Obtuse triangle

Acute triangle: All angles are less than 90 degrees

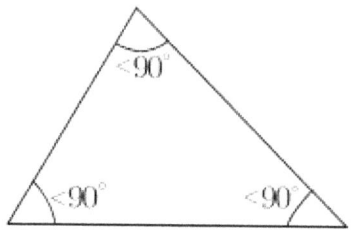

Acute triangle

1.1.2   Similarity and Congruence

Similar triangles: Similar triangles are same in shape but may not be same in size. Similar triangles are proportional triangles.

Two triangles are said to be similar, if every angle of one triangle has the same measure as the corresponding angle in the other triangle. The corresponding sides of similar triangles have lengths that are in the same proportion.

Necessary and sufficient conditions (Postulates) for similarity:

- Postulate PS-1.1.1: AA – If two angles of a triangle are same to two angles of another triangle.
- Postulate PS-1.1.2: SAS – If two sides of a triangle are in same proportion to another triangle, and their included angle is same
- Postulate PS-1.1.3: SSS – If all the three sides of a triangle are in same proportion to three sides of another triangle

Congruent triangles: Two triangles that are congruent have exactly the same size and shape: all pairs of corresponding interior angles are equal in measure, and all pairs of corresponding sides have the same length.

Necessary and sufficient conditions for congruence:

- Postulate PS-1.1.4: SAS - Two sides in a triangle have the same length as corresponding two sides in the other triangle, and the included angles of these two sides are also equal in two triangles.

- Postulate PS-1.1.5: ASA / AAS – Any two interior angles and one of any sides in a triangle have the same measure and length, respectively, as those in the corresponding angles and a side of other triangle.

- Postulate PS-1.1.6: SSS - Each side of a triangle has the same length as a corresponding side of the other triangle.

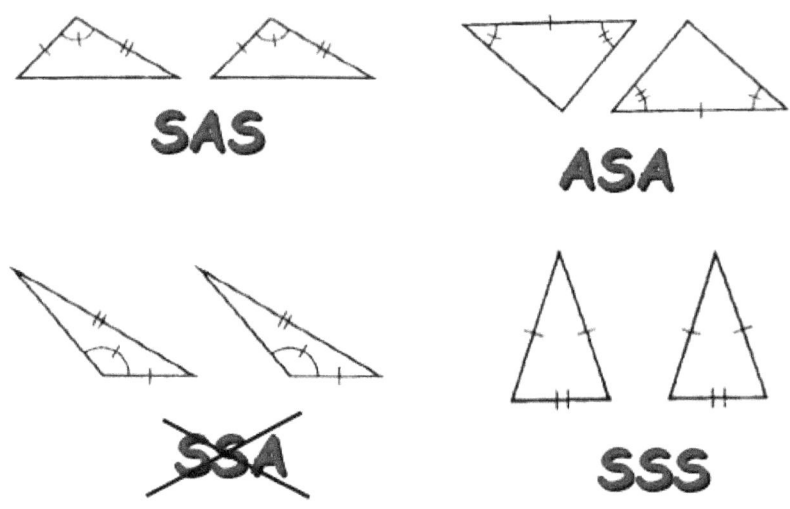

Conditions for Congruence

As depicted in figure above, the conditions for two triangles to be congruent are:-

a- Two sides and the included angle between them (SAS) should be equal for two triangles; *Note: Specifying two sides and an adjacent angle (SSA), however, can yield two distinct possible triangles.*

b- Two angles and the side between them (ASA), or two angles and any adjacent side (AAS) should be equal for two triangles;

c- All the three sides (SSS) should be equal for two triangles;

Worksheet-1

Exercise-1: Mention three postulates for similarity of triangles.

Exercise-2: Mention three postulates for congruence of triangles.

Exercise-3: The angles of a triangle always add up to 180°. Is this statement true or false?

Exercise-4: In the following figures from **4-a** to **4-c**, identify the triangle as Equilateral, Isosceles or Scalene.

4-a) if all angles are different.

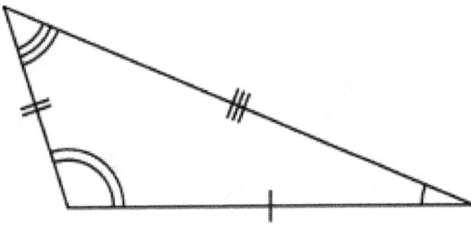

4-b) if the base angles are same.

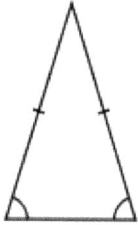

4-c) if all angles are equal.

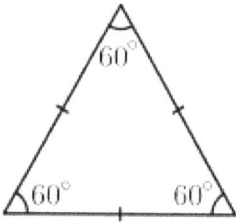

Exercise-5: For the figures depicted below from **5-a** to **5-c**, Identify the figure as acute, obtuse or right angle triangle:

5-a)

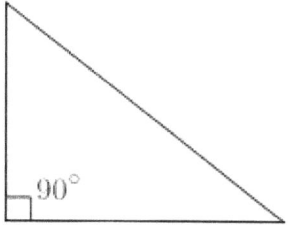

5-b)

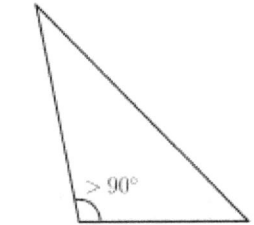

5-c)

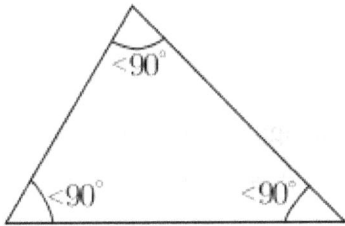

Exercise-6: For questions from **6-a** to **6-d**, check if two triangles are similar, congruent or possess no relation. Explain the postulate, if they are similar or congruent:

6-a) In the following figures of two triangles, angle B is equal to angle E and angle C is equal to F. None of the corresponding sides of two triangles are equal.

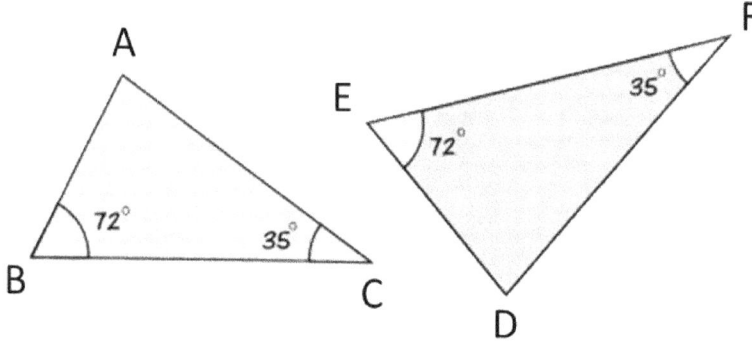

6-b) In the following figures of two triangles, all the angles and sides of a triangle are equal to corresponding angles and sides of another triangle.

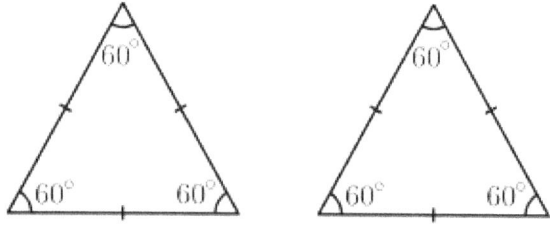

6-c) In the following figures of two triangles, angle B is equal to angle Y. None of corresponding sides of two triangles are equal.

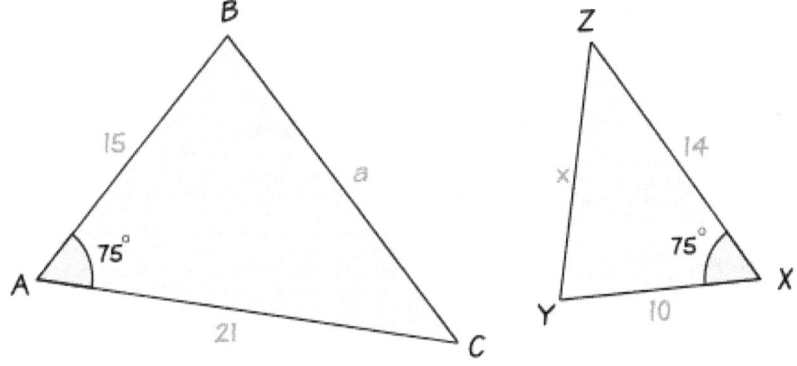

6-d) In the following figures of two triangles, angle A is same as angle D, angle B same as angle E, and side AC is same as side DF.

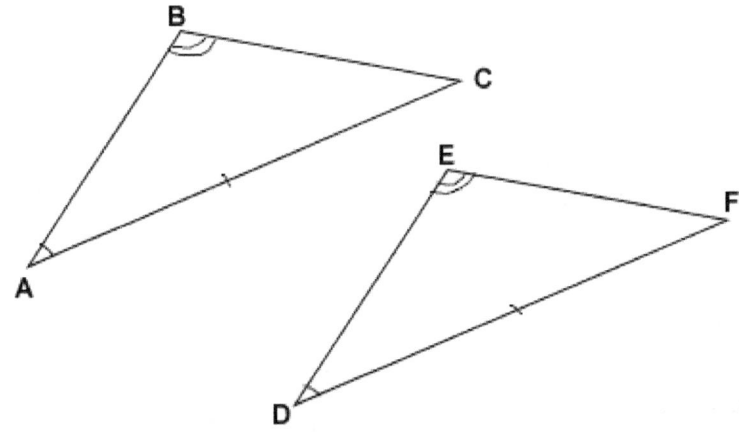

Solutions-Worksheet-1

Solution-1: AA, SAS, SSS; Corresponding sides proportional;

Refer Postulates PS-1.1.1, PS-1.1.2 and PS-1.1.3 of the chapter.

Solution-2: SAS (But not SSA/ASS), ASA / AAS, SSS; Corresponding sides equal;

Refer Postulates PS-1.1.4, PS-1.1.5 and PS-1.1.6 of the chapter.

Solution-3: True

Solution-4:

4-a) Scalene

4-b) Isosceles

4-c) Equilateral

Solution-5:

5-a) Right angle triangle

5-b) Obtuse angle triangle

5-c) Acute angle triangle

Solution-6:

6-a) As two angles of a triangle are equal to two angles of another triangle but no sides of two triangles are equal. The conditions are met to AA postulate of similarity. Therefore, two triangles are similar.

6-b) All the angles and sides of a triangle are same to corresponding angles and sides of another triangle. So conditions are met for all postulates of congruence (SAS, ASA, SSS). Therefore, two triangles are congruent

6-c) Angle B of a triangle equals angle Y of another triangle. Angle A of a triangle also equals angle X of another triangle. Sides BA and AC of a triangle are not same but in same proportion to YX and XZ of another triangle. Therefore AA postulate of similarity is fulfilled. Therefore, two triangles are similar.

6-d) Angle B of a triangle is equal to angle E of other triangle. Angle A of a triangle also equals angle D of other triangle. Side AC of a triangle is same to corresponding side DF of other triangle. Postulate AAS of congruence is fulfilled. Therefore, two triangles are congruent.

1.2   Basic Concepts and Theorems

<u>Right Angle</u>

A right angle equals 90 degrees.

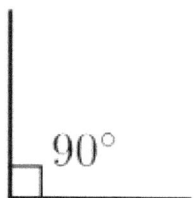

A right angle is an angle formed by two lines that are perpendicular to each other.

A straight line angle equals two right angles. Therefore, bisection of a straight line forms a right angle at both sides.

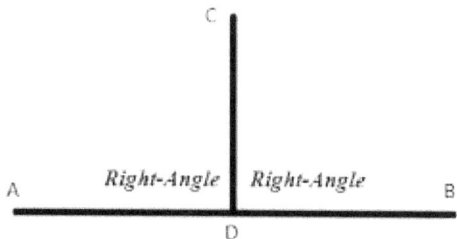

Alternatively, a right angle corresponds to a quarter turn of a full circle. Same way, two right angles correspond to a half turn of a full circle, which is an angle formed on a straight line.

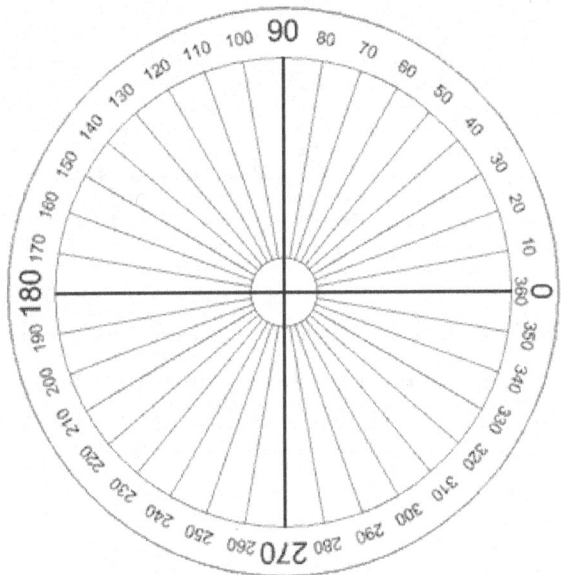

Full turn of a circle equals four right angles.

A straight line angle equals two right angles.

$$180°$$

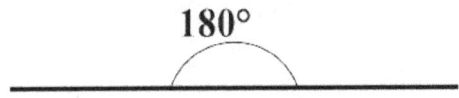

Postulate PS-1.2.7: if a straight line meets another straight line, then it makes two angles which are together equal to two right angles.

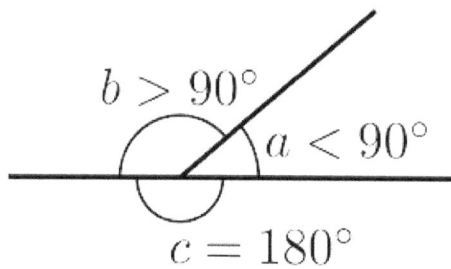

As per definition of right angle, a straight line angle equals two right angles. So a straight line meeting another straight line forms two angles, which sum together to two right angles.

Euclid's Elements

Euclid's Elements is a mathematical and geometric treatise consisting of 13 books written by Greek mathematician Euclid in 300 BC. It is a collection of postulates (axioms), propositions and mathematical proofs of theorems.

All of the postulates, theorems mentioned or derived here in this chapter are originally specified in Euclid's Elements: Book-1. We will not go deep into theoretical details but understand the main postulates and theorems which are important for us.

Terminology and Proposition for Angles of Transversal and Parallel Lines

Transversal is a line which intersects two or more parallel or non-parallel lines.

In the figure below, a transversal crosses two other parallel lines; it forms eight angles, as in our figure - angles 1, 2, 3, 4, 5, 6, 7 and 8.

In the figure below, line-1 and line-2 are parallel lines, which are intersected by a third straight line called transversal.

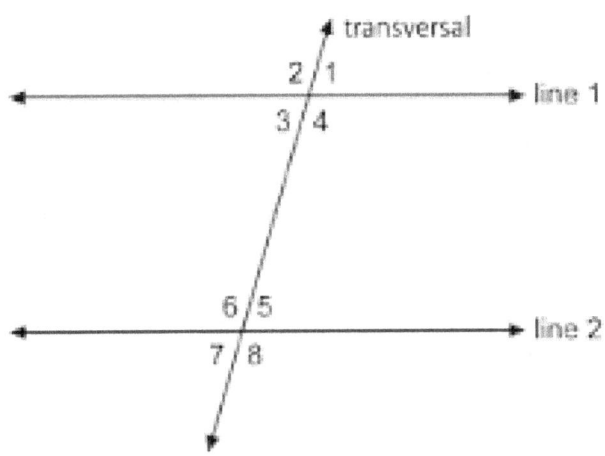

Terminology and Propositions for Angles of Transversal

Interior Angles are angles which are formed inside of two lines being intersected by transversal. In the figure, 3, 4, 5, and 6 are interior angles.

Exterior Angles are angles which are formed outside of two lines being intersected by transversal. In the figure, angles 1, 2, 7 and 8 are exterior angles.

Opposite Angles are a pair of nonadjacent angles formed by two intersecting lines. On line-1, 1 and 3 are opposite angles, as are angles 2 and 4. Opposite angles are always equal. Opposite angles are also called vertical angles.

Vertical Angles are a pair of nonadjacent angles formed by two intersecting lines. On line-1, 1 and 3 are vertical, as are angles 2 and 4. Vertical angles are always equal. So angle 1 equals 3 and similarly angle 2 equals 4.

Theorem T-1.2.1: *Vertical angles are congruent.*

Supplementary Angles are two angles formed on one side of a straight line that add up to 180 degrees. The sum of supplementary angles is always 180 degrees.

Supplementary angles of line-1 formed by transversal are: 2 and 1; as well as 3 and 4;

Supplementary angles at line-2 formed by transversal are: 6 and 5; as well as 7 and 8;

Theorem T-1.2.2: If two supplementary angles are formed on one side of a line, then these two angles form a linear pair, so they are supplementary.

Adjacent Angles are angles on same side of a straight line. Adjacent angles are supplementary, i.e. sum of two angles on a straight line is 180 degrees. Angles 2 and 3 are adjacent angles, as are 2 and 1; 5 and 8; and so on.

Theorem T-1.2.3: If two adjacent angles are on same side of a straight line, then the two angles form a linear pair, so they are supplementary.

Terminology / Propositions for Parallel Lines and Transversal

Corresponding Angles are considered to be in same location at each point of intersection. Angles 2 and 6 are corresponding angles because they both are located in the upper left corner. Similarly another pair of corresponding angles is 4 and 8, which are both in the lower right corner.

Condition of equality: If transversal crosses two parallel lines, i.e. if line-1 and line-2 are parallel, then corresponding angles are congruent (equal).

Theorem T-1.2.4: If two parallel lines are cut by a transversal, then the corresponding angles are congruent.

Theorem T-1.2.5: If two lines are cut by a transversal and the corresponding angles are congruent, then the lines are parallel.

Alternate Angles: Angles 5 and 3 are called alternate angles, as are angles 4 and 6. As the word Alternate explains; if we start at angle 5 and go around the interior angles, angle 3 alternates, so it is an alternate angle.

Alternate interior angles are congruent. Alternate exterior angles are congruent.

Refer theorem of Alternate Interior Angles.

Opposite & Adjacent Interior Angles

Any two interior angles on same side of transversal are called the adjacent interior angles. For parallel lines, the adjacent interior angles are supplementary. Angles 4 and 5 are adjacent interior angles and so are angles 3 and 6.

With respect to an exterior angle 1, angle 5 is the opposite interior angle. With respect to an exterior angle 1, angle 4 is an adjacent interior angle.

Theorem T-1.2.6: If two parallel lines are cut by a transversal, then the opposite interior angle with respect to any exterior angle are congruent.

Theorem T-1.2.7: If two parallel lines are cut by a transversal, then adjacent interior angle with respect to any exterior angle form a linear pair together, so they are supplementary.

Alternate Interior Angles are two interior angles which lie on different parallel lines and on opposite sides of a transversal.

In the given figure, angles 3 and 5 are alternate interior angles, as are angles 4 and 6.

Condition of equality: If transversal crosses two parallel lines, i.e. if line1 and line2 are parallel, then the alternate interior angles are congruent (equal).

Theorem T-1.2.8: If two parallel lines are cut by a transversal, then the alternate interior angles are congruent.

Theorem T-1.2.9: If two lines are cut by a transversal and the alternate interior angles are congruent, then the lines are parallel.

Alternate Exterior Angles are two exterior angles on opposite sides of a transversal which lie on different parallel lines. In the give figure, 2 and 8 are exterior angles, as are angles 1 and 7.

Condition of equality: If transversal crosses two parallel lines, i.e. if line1 and line2 are parallel, then the alternate exterior angles are equal.

Theorem T-1.2.10: If two parallel lines are cut by a

transversal, then the alternate exterior angles are congruent.

Theorem T-1.2.11: If two lines are cut by a transversal and the alternate exterior angles are congruent, then the lines are parallel.

Consecutive Interior Angles are two interior angles lying on the same side of the transversal cutting across two straight lines. So in our figure, 4 and 5 are consecutive interior angles, and so are 3 and 6.

Condition of supplementary: If transversal crosses two parallel lines, i.e. if line1 and line2 are parallel, then the consecutive interior angles are supplementary; i.e. they add up to 180 degrees.

Theorem T-1.2.12: If two parallel lines are cut by a transversal, then the interior angles on the same side of the transversal are supplementary.

Theorem T-1.2.13: If two lines are cut by a transversal and the interior angles on the same side of the transversal are supplementary, then the lines are parallel.

Consecutive Exterior Angles are two exterior angles lying on the same side of the transversal cutting across two straight lines. So in our figure, 1 and 8 are consecutive exterior angles, and so are 2 and 7.

Condition of supplementary: If transversal crosses two parallel lines, i.e. if line1 and line2 are parallel, then consecutive exterior angles are supplementary; i.e. they add up to 180 degrees.

Theorem T-1.2.14: **If two parallel lines are cut by a transversal, then the exterior angles on the same side of the transversal are supplementary.**

Theorem T-1.2.15: If two lines are cut by a transversal and the exterior angles on the same side of the transversal are supplementary, then the lines are parallel.

Some Important Propositions of Euclid's Elements:

Proposition P-1.2.1: If one side of a triangle is extended, then the exterior angle is greater than either of the opposite interior angles.

Proposition P-1.2.2: Any two angles of a triangle are together less than two right angles.

Proposition P-1.2.3: A greater side of a triangle is opposite a greater angle.

Proposition P-1.2.4: A greater angle of a triangle is opposite a greater side.

Proposition P-1.2.5: Any two sides of a triangle are together greater than the third side.

Theorems of Triangles

Theorem T-1.2.16: In any triangle the angle opposite the greater side is greater.

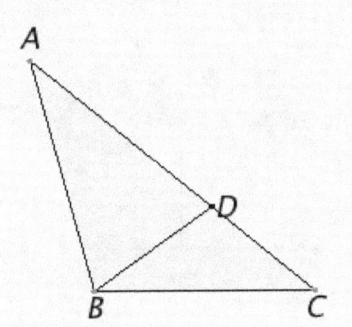

Proof: Since AC is greater than AB, a line BD is drawn to make AD equal to AB. So triangle ADB is isosceles triangle. Since the angle ADB is an exterior angle of the triangle BCD, therefore it is greater than the interior and opposite angle DCB.

As the triangle ADB is isosceles, the angle ADB equals the angle ABD, since the side AB equals AD. Therefore, the angle ABD is also greater than the angle ACB. Therefore the angle ABC is greater than the angle ACB.

Similarly, it is proved that the angle ABC is greater than the angle BAC.

Therefore in any triangle the angle opposite the greater side is greater.

Conversely, it is also true that a side opposite to a greater angle is greater.

Remarks: Six postulates for similarity and congruence of triangles, discussed earlier are important to prove some basic theorems.

Theorem T-1.2.17: If a straight line bisects the vertex angle of an isosceles triangle, then the line is perpendicular bisector of base.

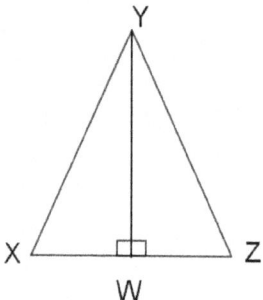

Proof: Triangles XYW, ZYW are congruent due to SAS postulate PS-1.1.4, because:

Side XY is same as YZ because XYZ is an isosceles triangle. Moreover, angles XYW and ZYW are same because YW bisects the vertex angle XYZ. Additionally, side YW is common to both the triangles.

Therefore, angle XWY is equal to angle ZWY. As the base angle XWZ is 180 degrees (two right angles) and is divided equally in two parts between angles XWY and ZWY. Therefore XWY and ZWY are each right angles. YW therefore is the perpendicular bisector of the base.

Theorem T-1.2.18: In an isosceles triangle, the angles at the base are equal.

Proof: As we have proved in theorem above, if a line bisects the vertex angle of an isosceles triangle and meets the base forming two triangles, then these two triangles are congruent with each other.

It is hence proved that the angles at the base of an isosceles triangle are equal.

Conversely, if two angles of a triangle are equal, then the sides opposite them will be equal. Refer ASA/AAS postulate PS-1.1.5.

Worksheet-2

Exercise-1: Referring the figure below where line-1 and line-2 are parallel intersected by a transversal, solve the exercises from 1-1 to 1-8.

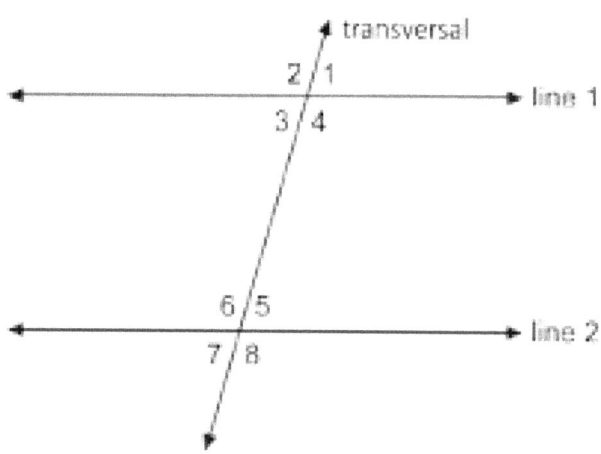

Exercise 1-1: Identify the following:

a) Interior Angles
b) Exterior Angles
c) Opposite Angles
d) Vertical Angles
e) Supplementary Angles
f) Adjacent Angles
g) Corresponding Angles
h) Alternate Angles
i) Alternate Interior Angles
j) Alternate Exterior Angles
k) Consecutive Interior Angles
l) Adjacent Interior Angles
m) Consecutive Exterior Angles

n) With respect to exterior angle 1, identify the opposite interior angle.

o) With respect to exterior angle 1, identify the adjacent interior angle.

Exercise-1-2: Which vertical angles are equal?

Exercise-1-3: Which angles in the figure are supplementary?

Exercise-1-4: Which corresponding angles are congruent?

Exercise-1-5: Which alternate interior angles are congruent?

Exercise-1-6: Which alternate exterior angles are congruent?

Exercise-1-7: Which consecutive interior angles are supplementary?

Exercise-1-8: Which consecutive exterior angles are supplementary?

Exercise-2: Prove the theorem:

In any triangle the angle opposite the greater side is greater.

Exercise-3: Prove the theorem:

If a straight line bisects the vertex angle of an isosceles triangle, then the line is perpendicular bisector of base.

Exercise-4: Prove the theorem:

In an isosceles triangle, the angles at the base are equal.

Solutions-Worksheet-2

Solutions-1-1:

a) 3, 4, 5, and 6 are interiors angles.

b) 1, 2, 7 and 8 are exterior angles.

c) On line-1, 1 and 3; 2 and 4; are opposite angles.
   On line-2, 6 and 8; 5 and 7; are opposite angles.

d) On line-1, 1 and 3; 2 and 4; are vertical angles.
   On line-2, 6 and 8; 5 and 7; are vertical angles.

e) On line-1, 1 and 2; 3 and 4; are supplementary angles.
   On line-2, 5 and 6; 7 and 8; are supplementary angles.

f) On line-1, 1 and 2; 3 and 4; are adjacent angles.
   On line-2, 5 and 6; 7 and 8; are adjacent angles.

g) 2 and 6; 3 and 7; 1 and 5; 4 and 8; are corresponding angles.

h) 5 and 3; 4 and 6; are alternate angles.

i) 5 and 3; 4 and 6; are alternate interior angles.

j) 2 and 8; 1 and 7; are alternate exterior angles.

k) 4 and 5; 3 and 6; are consecutive interior angles

l) 4 and 5; 3 and 6; are adjacent interior angles

m) 1 and 8; 2 and 7; are consecutive exterior angles

n) With respect to exterior angle 1, angle 5 is the opposite interior angle.

o) With respect to exterior angle 1, angle 4 is the adjacent interior angle.

Solution-1-2: On line-1, 1 and 3; 2 and 4; are vertical angles.

On line-2, 6 and 8; 5 and 7; are vertical angles.

Solution-1-3: Adjacent angles are supplementary.

On line-1, 1 and 2; 3 and 4; are adjacent angles.

On line-2, 5 and 6; 7 and 8; are adjacent angles.

Solution-1-4: 2 and 6; 3 and 7; 1 and 5; 4 and 8; are congruent corresponding angles because line-1 and line-2 are parallel.

Solution-1-5: 5 and 3; 4 and 6; are congruent alternate interior angles because line-1 and line-2 are parallel.

Solution-1-6: 2 and 8; 1 and 7; are congruent alternate exterior angles because line-1 and line-2 are parallel.

Solution-1-7: 4 and 5; 3 and 6; are supplementary consecutive interior angles.

Solution-1-8: 1 and 8; 2 and 7; are supplementary consecutive exterior angles.

Solution-2: Refer Theorem T-1.2.16 in the section 1.2 of this chapter.

Solution-3: Refer Theorem T-1.2.17 in the section 1.2 of this chapter.

Solution-4: Refer Theorem T-1.2.18 in the section 1.2 of this chapter.

1.3   A Right Triangle

Most essential triangle in our discussion is a right triangle. A right triangle is the one in which one of the angles is 90°. Let's look at right triangle.

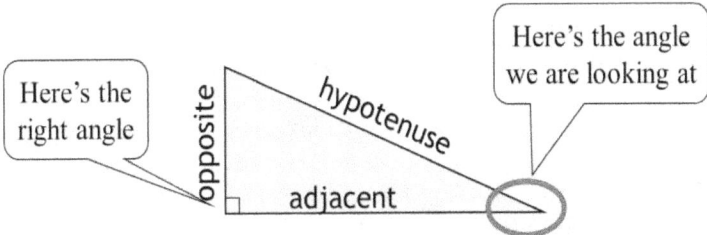

- In a right triangle, we pick one of the acute angles other than right angle, as marked in circle. We name the other two sides relative to the marked angle.

- We call the longest side the hypotenuse (hyp), which is opposite to right angle.

- The side opposite to selected acute angle is called opposite (opp) or perpendicular.

- The side between the selected acute angle and right angle is called adjacent (adj) or base.

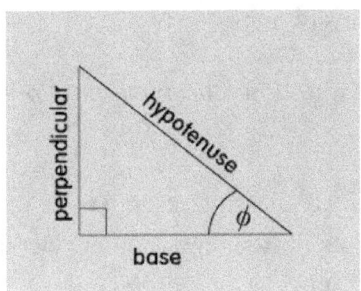

1.3.1   The Pythagorean Theorem

The Pythagorean Theorem / Pythagoras's theorem - states that the sum of the squares of the legs of a right triangle will equal the square of the hypotenuse of the triangle.

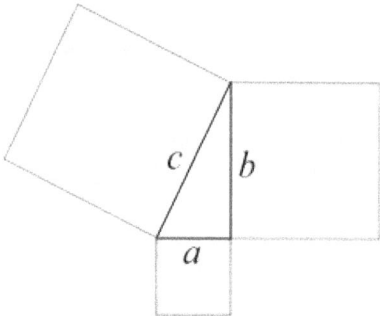

The sum of the areas of the two squares on the legs (*a* and *b*) equals the area of the square on the hypotenuse (*c*) in any right triangle.

$$a^2 + b^2 = c^2,$$

Euclid's elements mention an important proposition on Pythagorean Theorem as follows:

Proposition P-1.3.1: In a right triangle, the square drawn on the side opposite the right angle is equal to the squares drawn on the sides that make the right angle.

Conversely, we have another proposition from Euclid's Elements as follows:

Proposition P-1.3.2: If the square drawn on one side of a triangle is equal to the squares drawn on the other two sides, then the angle contained by those two sides is a right angle.

Let's look into squares better with grid.

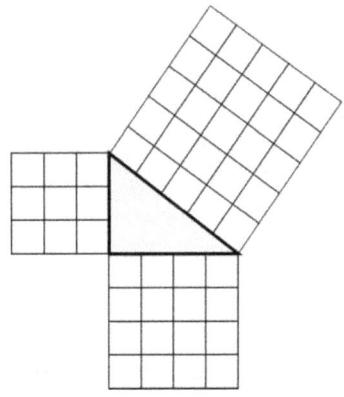

The area of square shows:

$$base^2 + perpendicular^2 = hypotenuse^2$$

Example-1: Find the distance between the points (−3, 2) and (2, 5) using Pythagorean Theorem.

Plotting these points on the coordinate plane, we draw a right triangle with the given points at each end of the hypotenuse.

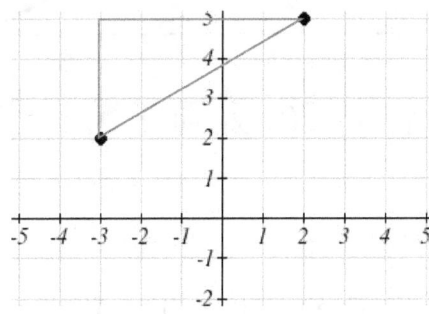

We find horizontal width of the triangle to be 5 and the vertical height to be 3. Finally, we calculate the distance between the points using the Pythagorean Theorem as follows:

$$dist^2 = 5^2 + 3^2 = 34$$
$$dist = \sqrt{34}$$

Distance Formula

We can apply Pythagorean Theorem to find the distance between two points in a coordinate plane.

The distance between two points (x_1, y_1) and (x_2, y_2) can be calculated as:

$$dist = \sqrt{(x_2 - x_1)^2 + (y_2 - y_1)^2}$$

Consider a triangle having ABC as given in figure:

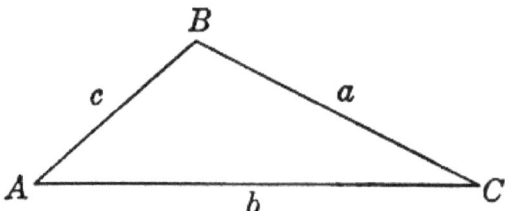

Considering the points A (x_1, y_1) and B (x_2, y_2), the distance AB is given by the formula:

Distance AB $= \sqrt{(x_2 - x_1)^2 + (y_2 - y_1)^2}$

We can use the distance formula to find the distance between each pair of points making up our triangle. These distances are the lengths of the three sides.

Example-2: For the three points (1, 3), (–2, –2) and (3, –1), on a coordinate plane, calculate the sides of a triangle.

On a coordinate plane, three points on the (x, y) plane are plotted and joined together with lines.

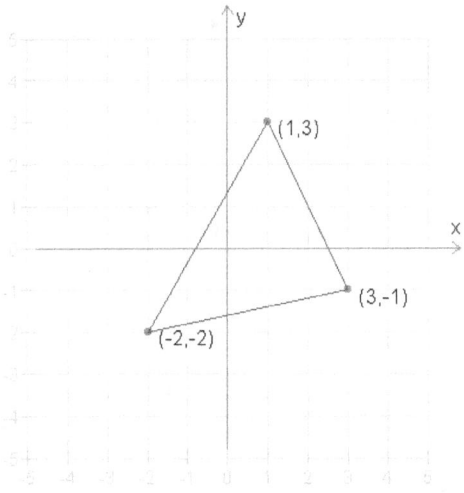

If the points are not in the same straight line, then a triangle is created. This diagram shows the triangle created by the three points (1, 3), (–2, –2) and (3, –1).

We find the magnitude of a side joining (1, 3) and (–2, –2)

$$\textbf{Distance} = \sqrt{(x_2 - x_1)^2 + (y_2 - y_1)^2}$$
$$= \sqrt{(-2 - 1)^2 + (-2 - 3)^2}$$
$$= \sqrt{9 + 25} = \sqrt{34} = 5.8309$$

Similarly, the side joining (1, 3) and (3, −1) is 4.47214, and the side joining (−2, −2) and (3, −1) is 5.09902

We have calculated side lengths of the triangle, as shown below:

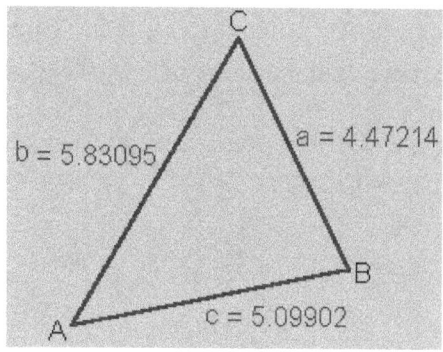

Right Triangle on a Coordinate Plane

Right angle aligned to X-Y axis: Let the right triangle be ABC with the vertices A, B and C, where the base and perpendicular are parallel to X-Y axis in any order. In such case, just looking at the coordinate points, we can identify the vertex of right angle.

In case of right triangle with base/perpendicular parallel to X-Y axis, a vertex at right angle must have the X-coordinate same as X-coordinate of any one end of hypotenuse; and the Y-coordinate same as Y-coordinate of other end of hypotenuse.

Example-1: In a right triangle with vertices A (4, −5), B (4, 4) and C (3, 4), base is parallel to X-axis. Find out the points making the hypotenuse.

Solution: In given right triangle, only point B has X-coordinate as 4, which is also X-coordinate of A. Moreover, Point B also has Y-coordinate as 4, which is Y-coordinate of C.

Therefore, point B is at right angle and side AC makes the hypotenuse.

<u>Non-aligned Right Triangle on a Coordinate Plane</u>:

For right triangle on a coordinate plane, where base or perpendicular is not parallel to X-Y axis, we can't identify the vertex of right angle only by looking at the coordinate points. We will have to apply distance formula to find sides and then using Pythagorean Theorem we can find the vertex of right angle.

<u>Example-2</u>: In a right triangle with vertices A (5, –3), B (2, –6) and C (1, 1). Find out the points making the hypotenuse.

Solution: We will apply distance formula to find the sides.

$$AB = \sqrt{(2-5)^2 + (-6-(-3))^2}$$
$$= \sqrt{9+9} = 4.243$$
$$BC = \sqrt{(1-2)^2 + (1-(-6))^2}$$
$$= \sqrt{1+49} = 7.071$$
$$CA = \sqrt{(5-1)^2 + (-3-1)^2}$$
$$= \sqrt{16+16} = 5.657$$

Applying Pythagorean Theorem, we find the relation as

$$\boldsymbol{AB^2 + CA^2 = BC^2}$$

Therefore, BC is the hypotenuse and vertex A is at right angle.

Worksheet-3

Exercise-1: In following triangle, identify the hypotenuse, perpendicular, and base with respect to angle C (∠ BCA).

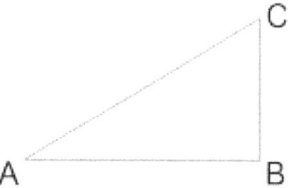

Exercise-2: In the following right triangle, find the size of hypotenuse.

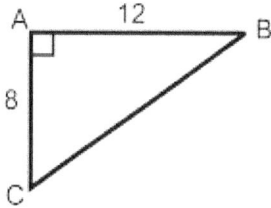

Exercise-3: The distance formula can be used to find the distance between two coordinate points. – True/False

Exercise-4: Explain the distance formula.

Exercise-5: Find the possible third points of right triangle with base parallel to X-axis, if the two vertices of hypotenuse on coordinate plane are (–3, 2) and (2, 6).

Exercise-6: Find the distance between the points (–3, 2) and (2, 6) using pythagorean theorem.

Exercise-7: Three points (3, –2), (–2, –2) and (3, 4) on a coordinate plane joined together, calculate the sides of the triangle.

Exercise-8: Given three coordinate points (3, –5), (–4, –5) and (3, 4) as vertices of a right triangle, where perpendicular is parallel to Y-axis. Find out which two points make the hypotenuse of this triangle without actually plotting on a coordinate plane.

Exercise-9: In a right triangle with vertices A (5, –3), B (2, –6) and C (1, 1). Find out the points making the hypotenuse.

Solutions-Worksheet-3

Solution-1: With respect to angle C, hypotenuse is AC; perpendicular is AB; and base is BC.

Solution-2: We apply the Pythagorean Theorem to find hypotenuse BC.

$$hypotenuse^2 = base^2 + perpendicular^2$$

$$BC^2 = AC^2 + AB^2 = 8^2 + 12^2 = 64 + 144 = 208$$

$$BC = 14.42$$

Solution-3: True

Solution-4: Considering two coordinate points A (x_1, y_1) and B (x_2, y_2), the distance AB is given by the distance formula as:

$$Distance = \sqrt{(x_2 - x_1)^2 + (y_2 - y_1)^2}$$

Solution-5: For two coordinate points (x_1, y_1) and (x_2, y_2) to be a hypotenuse of a right triangle with base parallel to X-axis, there are two possible points which could be a third point of this right triangle, (x_1, y_2) and (x_2, y_1)

If two end points of hypotenuse are (–3, 2) and (2, 6), then the third point is either (–3, 6) or (2, 2).

Solution-6: As per distance formula

$$Distance = \sqrt{(x_2 - x_1)^2 + (y_2 - y_1)^2}$$

$$= \sqrt{(2 - (-3))^2 + (6 - 2)^2}$$

$$= \sqrt{25 + 16} = \sqrt{41} = 6.403$$

Solution-7: Let the right triangle be ABC with the vertices A (3, – 2); B (–2, –2); and C (3, 4);

$$AB = \sqrt{(-2-3)^2 + (-2-(-2))^2}$$

$$= \sqrt{25 + 0} = 5$$

$$BC = \sqrt{(3-(-2))^2 + (4-(-2))^2}$$

$$= \sqrt{25 + 36} = \sqrt{61} = 7.81$$

$$CA = \sqrt{(3-(3))^2 + (4-(-2))^2}$$

$$= \sqrt{0 + 36} = 6$$

Solution-8: Let the right triangle be ABC with the vertices A (3, – 5), B (–4, –5) and C (3, 4). As the perpendicular is parallel to Y-axis in this right triangle, a vertex at right angle must have the X-coordinate same as X-coordinate of any one end of hypotenuse; and the Y-coordinate same as Y-coordinate of other end of hypotenuse.

In given right triangle, only point A has X-coordinate as 3, which is also X-coordinate of C; and the Y-coordinate as –5, which is also Y-coordinate of B.

Therefore, point A is the vertex at right angle and side BC makes the hypotenuse.

Solution-9: We will apply distance formula to find the sides.

$$AB = \sqrt{(2-5)^2 + (-6-(-3))^2}$$

$$= \sqrt{9 + 9} = 4.243$$

$$BC = \sqrt{(1-2)^2 + (1-(-6))^2}$$

$$= \sqrt{1+49} = 7.071$$

$$CA = \sqrt{(5-1)^2 + (-3-1)^2}$$

$$= \sqrt{16+16} = 5.657$$

Applying pythagorean theorem, we find $AB^2 + CA^2 = BC^2$ Therefore, BC is the hypotenuse and vertex A is at right angle.

2 Chapter 2: Angles and Trigonometric Ratios

2.1　Angle and its Measurement

An angle is a measure of rotation of a given line about its initial side. The initial position of line is initial side; and the final position after rotation is terminal side of angle. The point of rotation is vertex.

If the direction of rotation is anticlockwise, the angle is assumed as positive. If the direction of rotation is clockwise, the angle is said to be negative.

The symbol $\angle$ is used to denote the angle.

Using Greek Symbols for angles

When representing angles using variables, it is traditional to use Greek letters.　Here is a list of commonly encountered Greek letters.

θ	φ or ϕ	α	β	γ
Theta	phi	alpha	beta	gamma

Positive Angle: The direction of rotation is anticlockwise.

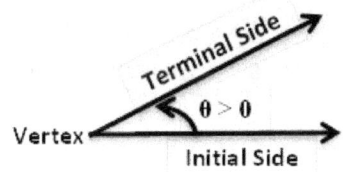

Negative Angle: The direction of rotation is clockwise.

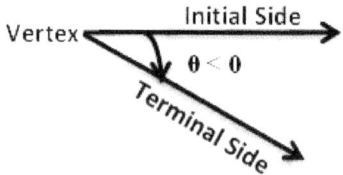

Arc length

Arc-length is the length of an arc, s, along a circle of radius r subtended (drawn out) by an angle θ. It is the portion of the circumference between the initial and terminal sides of the angle.

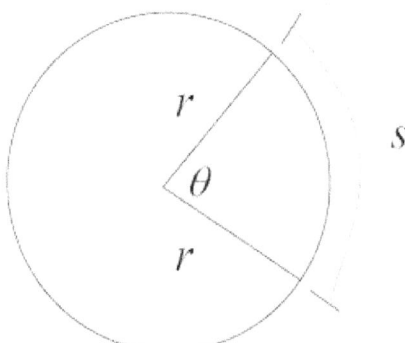

Angle in Radian

Radian, commonly used as a measure of angle, is based on the distance around a circle.

The radian measure of an angle is the ratio of the length of the circular arc subtended by the angle to the radius of the circle.

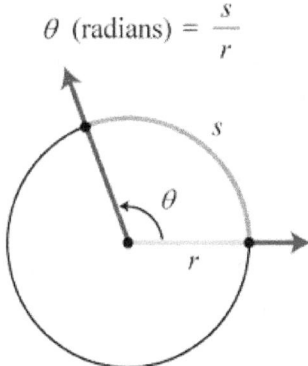

$$\theta \text{ (radians)} = \frac{s}{r}$$

In other words, if s is the length of an arc of a circle, and r is the radius of the circle, then

$$\text{radian measure} = \frac{s}{r}$$

Properties of radian with respect to a circle:

a- If the circle has radius 1, then the radian measure corresponds to the length of the arc.

b- For angle of 1 radian, the arc length equals radius.

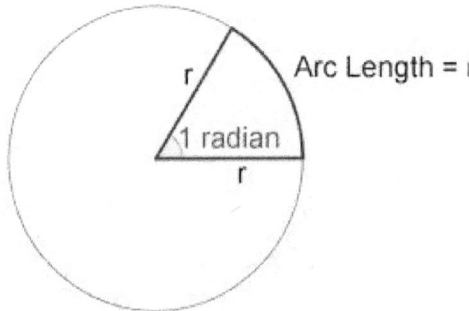

c- One complete angle of any circle is 2π.

d- The length of the arc around an entire circle is called the circumference of a circle. The circumference C of a circle is formulated as: $C = 2\,\pi\,r$

e- The ratio of the circumference to the radius produces the constant: $2\,\pi$. Regardless of the radius, this ratio is always the same, just as how the degree measure of an angle is independent of the radius.

Radian is a pure number and has no unit. So, when an angle is expressed in radian measure, the word radian is often omitted.

Angle in Degrees

Rotation of $(\frac{1}{360})th$ of a revolution from initial side to terminal side is said to be one degree. A degree is divided into 60 minutes, and a minute is divided into 60 seconds. A degree is written as 1°, a minute as $1'$, and a second as $1''$.

$$1^\circ = 60' \; ; \quad 1' = 60'' \; ; \quad 1^\circ = 3600'' \; ;$$

Some of the common angles in degrees are shown below:-

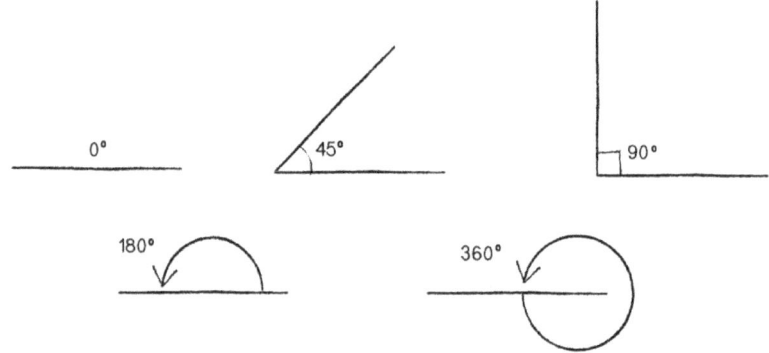

If rotation from initial side is more than one revolution, then angles is more than 360 degrees, as shown in figure for 405 degrees. If the rotation is in clockwise direction, the angle is negative as shown in figure.

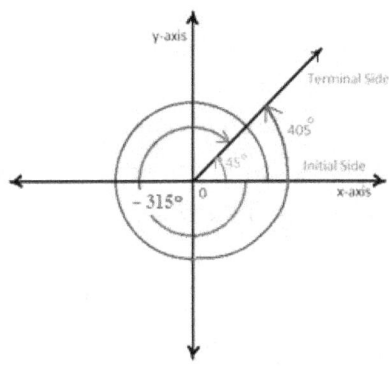

Angles in Degrees and Radians

Simplest way of relating degree with radian is the angle of a complete revolution of circle. Radian measure of a circle is 2π and degree measure is 360 degrees. So, 2π radian = 360°. Similarly, π radian = 180°.

The relation between degree and radian of some common angles are:

Degree	30°	45°	60°	90°	180°	270°	360°
Radian	$\pi/6$	$\pi/4$	$\pi/3$	$\pi/2$	π	$3\pi/2$	2π

Exchanging radian and degree:

1 radian = 180° / π = 57° 16″ approximately.

1° = π / 180 radian = 0.01746 radian approximately.

As explained, angles are measured in terms of degrees and radians. We can easily convert the radian into degrees and vice versa using following formula.

$$\frac{R}{\pi} = \frac{D}{180°}$$, where R is radian and D is degree measure

Alternatively, to convert degrees into radian and vice versa, we can use per-unit method using: π *radian* $= 180°$. *(We will use this unit method very often in this book.)*

Example-1: Convert 45° into radian.

Solution: Substituting as per formula: $\frac{R}{\pi} = \frac{D}{180°}$, we have

$$R = \frac{45° \ \pi}{180°} = \frac{\pi}{4}$$

Example-2: Convert $\frac{\pi}{3}$ into degrees.

Solution: Substituting as per formula: $\frac{R}{\pi} = \frac{D}{180°}$, we have

$$D = \frac{180°}{\pi} \ \frac{\pi}{3} = 60°$$

Example-3: Find the arc-length along a circle of radius 10 cm subtended by an angle of 225 degrees.

Solution: Given quantities are angle as 225 degrees and radius r as 10 cm. Assuming $\pi = 3.143$, angle 225 degrees is converted into radian as,

$$\text{Angle in radian} = \frac{225° \; \pi}{180°} = 1.25 \, \pi = 3.93 \text{ (approx.)}$$

We find arc-length using formula,

Arc length = Angle in radian x Radius

Substituting the values, we get arc length = 3.93 x 10 = 39.3 cm

---

$\pi/3$ = 60 degrees; π = 180 degrees; 2π = 360 degrees;

Note: Radian is a pure number and has no unit. π is a constant approximated as 3.1416.

Coterminal Angles

There are 360 degrees in one rotation, an angle greater than 360 degrees would indicate more than 1 full rotation.

Two angles are said to be coterminal, if they terminate at the same position, so their terminal sides coincide. We can find coterminal angles by adding or subtracting a multiple of full rotation (*n x 360 degrees*).

For example, an angle of 390 degrees is coterminal with an angle of 390–360 = 30 degrees.

As another example, an angle of 800 degrees is coterminal with an angle of 800–360 = 440 degrees. It would also be coterminal with an angle of 440–360 = 80 degrees. So, $\theta = 80°$ is coterminal with 800°.

Similarly, for – 45° a positive coterminal angle is found by adding

360 degrees: $-45° + 360° = 315°$. So, $-45°$ is coterminal with $315°$.

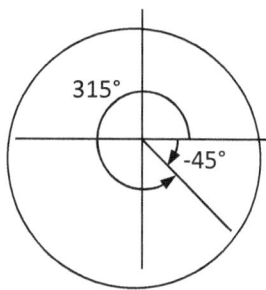

When working in degrees, we find coterminal angles by adding or subtracting a multiple of 360 degrees (full rotation). Likewise, in radians, we can find coterminal angles by adding or subtracting a multiple of 2π radians.

As an example, we find coterminal angle for $\dfrac{19\pi}{4}$

$$\frac{19\pi}{4} - 2\pi = \frac{19\pi}{4} - \frac{8\pi}{4} = \frac{11\pi}{4}$$

The angle $\dfrac{11\pi}{4}$ is coterminal, but not less than 2π, so we subtract another rotation.

$$\frac{11\pi}{4} - 2\pi = \frac{11\pi}{4} - \frac{8\pi}{4} = \frac{3\pi}{4}$$

Hence, the angle $\dfrac{3\pi}{4}$ is coterminal with $\dfrac{19\pi}{4}$.

Worksheet-4

[Use π = 3.143]

Exercise-1: Convert 1 radian into degree

Exercise-2: Convert 1 degree into radian

Exercise-3: Convert 0.01746 radian into degree

Exercise-4: Convert 57°16″ into radian

Exercise-5: Convert 7 radians into degree

Exercise-6: Find the radian measure for angle $-47^\circ\ 30'$.

Exercise-7: Convert 520° into radian.

Exercise-8: Find the degree measure of 11/16 radian.

Exercise-9: Convert –4 radians into degree.

Exercise-10: Convert ($7\pi / 6$) radian into degree.

Exercise-11: A wheel makes 180 revolutions in one minute. Through how many radians does it turn in one second?

Exercise-12: Find the degree measure of angle subtended at the centre of circle of radius 10 cm by an arc of 11 cm.

Exercise-13: If in two circles, arcs of same length subtend angles 60° and 90° at the centre, find the ratio of their radii.

Exercise-14: Find the angle in radian through which a pendulum of length 25 cm swings the arc of length 15 cm.

Exercise-15: The minute hand of a watch is 1.5 cm long. In 45 minutes, how long does its tip move?

Exercise-16: Find the radius of a circle in which central angle of 45 degree touches the arc of length 7 cm.

Exercise-17: Find the radius of circle which intercepts the arc of length 14 cm making a central angle of 60 degree.

Exercise-18: The angles subtended by arcs in a circle are 30° and 45°. Find the ratio of arcs.

Exercise-19: Find an angle α that is coterminal with 870°, where $0^\circ \leq \alpha < 360^\circ$.

Exercise-20: Find an angle β coterminal with −300°, where 0° ≤ β < 360°.

Exercise-21: Find an angle β that is coterminal with $-\dfrac{17\pi}{6}$, where $0 \le \beta < 2\pi$

Exercise-22: Find an angle β that is coterminal with $\dfrac{19\pi}{4}$, where $0 \le \beta < 2\pi$

Solutions-Worksheet-4

Solution-1:

π radian = 180°

1 radian = 180 / π degree = 180 / 3.143 degree = 57.27°

Converting the fractional part of degree into minute:

1° = 60′ , so 0.27° = 16.2′

Hence, 1 radian = 57°16′ (approx.)

Solution-2:

180° = π radian

So, 1° = π / 180 radian = 3.143 / 180

= 0.01746 radian approx.

Solution-3:

π radian = 180°

So, 0.01746 radian = (180 x 0.01746) / 3.143 = 1° approx.

Solution-4:

First converting the 16 seconds into degree:

3600″ = 1° , so 16″ = (1 x 16) / 3600 degree = 0.0044° approx.

Now, we need to convert 57.0045° into radian,

$$180° = \pi \text{ radian}$$

So, 57.0045° = (3.143 x 57.0045) / 180 radian

$$= 0.9954 \text{ radian approx.}$$

Solution-5:

$$\pi \text{ radian} = 180°$$

So, 7 radian = (180 x 7) / 3.143 degree = 400.89° approx.

Converting the fractional part of degree into minutes:

$$1° = 60', \text{ therefore } .89° = 53.4'$$

Converting the fractional part of minute into seconds:

$$1' = 60'', \text{ therefore } 0.4' = 24''$$

Therefore, 7 radian = 400° 53′ 24″

Solution-6: –47° 30′ = –47.5°

$$180° = \pi \text{ radian}$$

So, –47.5° = (3.143 x –47.5) / 180 = –0.829 radian

Solution-7:

$$180° = \pi \text{ radian}$$

So, $520° = (3.143 \times 520) / 180 = 9.08$ radian

Solution-8:

π radian $= 180°$

So, $11/16$ radian $= (180 /3.143) \times (11/16) = 57.27 \times 0.6875 = 39.37°$

Converting the 0.37 degrees into minutes,

$1° = 60'$, so $0.37° = 22.2'$

Converting the 0.2 minutes into seconds,

$1' = 60''$, so $0.2' = 12''$

Therefore, $11/16$ radian $= 39°22'12''$

Solution-9:

π radian $= 180°$

So, -4 radian $= (180 /3.143) \times (-4) = -229.08°$

Converting the 0.08 degrees into minutes,

$1° = 60'$, so $0.08° = 4.8'$

Converting the 0.8 minutes into seconds,

$1' = 60''$, so $0.8' = 48''$

Therefore, -4 radian $= -229.08° \ 4' \ 48''$

Solution-10:

π radian = 180°

So, $7\pi/6 = (180/\pi) \times (7\pi/6) = 210°$

Solution-11: In 60 seconds, the wheel turns $180 \times (2\pi)$ radian.

So in 1 second, it turns $180 \times (2\pi) / 60 = 6\pi$ radian

Solution-12: Given quantities are arc s as 11 cm and radius r as 10 cm.

Angle in radian = $\dfrac{arc}{radius} = \dfrac{11}{10} = 1.1$ radian

Converting radian into degrees, $(180 \times 1.1)/3.143 = 62.99°$

Solution-13: Let the radii be r1 (bigger radius) and r2 (smaller radius) and equal arcs be s.

Angles are converted into radians, $60° = \pi/3$; and $90° = \pi/2$;

As per definition, arc = radius x angle

For circle of bigger radius, $s = r1 \times \pi/3$ -------Eq(1)

For circle of smaller radius, $s = r2 \times \pi/2$ -------Eq(2)

Equating the Eq(1) and Eq(2) together, we have

$r2 \times \pi/2 = r1 \times \pi/3$

Finding the ratio of circles (smaller to bigger radius):

$$\frac{r2}{r1} = \frac{\pi/3}{\pi/2} = \frac{2}{3}$$

Solution-14: As per definition, angle in radian = $\dfrac{arc}{radius}$

The length of pendulum 25 cm is radius and arc is 15 cm. So,

$$\text{angle} = \frac{15}{25} = \frac{3}{5} \text{ radian}$$

Solution-15: As per definition, arc = angle (in radian) x radius

In 60 minutes, the minute hand moves one complete revolution i.e. 2π. In 45 minutes, the minute hand moves $3\pi/2$.

The length of minute hand 1.5 cm is the radius. So,

$$\text{arc} = (3\pi/2) \text{ x } 1.5 = 4.714 \text{ x } 1.5 = 7.071 \text{ cm (approx.)}$$

Solution-16:

As per definition, radius = arc / angle (in radian)

Converting 45 degree in radian, $(\pi/180)$ x 45 = 0.786 radian

radius = 7 / 0.786 = 8.906 cm (approx.)

Solution-17: As per definition, radius = arc / angle (in radian)

Converting 60 degree in radian, $(\pi/180)$ x 60 = 1.048 radian

radius = 14 / 1.048 = 13.36 cm (approx.)

Solution-18: Let the radii be r1 (bigger radius), r2 (smaller radius), and equal arcs be s.

Angles are converted into radians, $30° = \pi/6$; and $45° = \pi/4$;

As per definition, arc = radius x angle

For circle of bigger radius, $s = r1 \times \pi/6$ -------Eq(1)

For circle of smaller radius, $s = r2 \times \pi/4$ -------Eq(2)

Equating the Eq(1) and Eq(2) together, we have

   $r2 \times \pi/4 = r1 \times \pi/6$

Finding the ratio of circles (smaller to bigger radius):

$$\frac{r2}{r1} = \frac{\pi/6}{\pi/4} = \frac{4}{6} = \frac{2}{3}$$

Solution-19: We find coterminal angles by adding or subtracting multiple of a full rotation (360 degrees).   Subtracting two full rotations, we get

   $870° - 2\,(360°) = 150°$

As $0° \leq 150° < 360°$, 150° is coterminal with 870°.

Solution-20: We find coterminal angles by adding or subtracting multiple of a full rotation (360 degrees). Adding a full rotation,

   $-300° + 360° = 60°$

As $0° \leq 60° < 360°$, 60° is coterminal with –300°.

Solution-21: We can find coterminal angles by adding or subtracting some full rotations of 2π radians. Adding two rotations, we have,

$$-\frac{17\pi}{6} + 2\,(2\pi) = \frac{7\pi}{6}$$

As $0 \le \frac{7\pi}{6} < 2\pi$, so $\frac{7\pi}{6}$ is the coterminal angle with $-\frac{17\pi}{6}$.

Solution-22: We can find coterminal angles by adding or subtracting some full rotations of 2π radians.

$$\frac{19\pi}{4} - 2\pi = \frac{19\pi}{4} - \frac{8\pi}{4} = \frac{11\pi}{4}$$

The angle $\frac{11\pi}{4}$ is coterminal, but not less than 2π, so we subtract another rotation.

$$\frac{11\pi}{4} - 2\pi = \frac{11\pi}{4} - \frac{8\pi}{4} = \frac{3\pi}{4}$$

Hence, the angle $\frac{3\pi}{4}$ is coterminal with $\frac{19\pi}{4}$.

2.2   Trigonometric Ratios – Sin, Cos, Tan

Trigonometric ratios are considered for acute angles as the ratio of sides of right angled triangle. Since a triangle has three sides, there are six ways to divide the lengths of the sides. Each of these six ratios has a name (and an abbreviation).

We consider a right angled triangle ABC as depicted below. For acute angle A, side a is perpendicular, side b is base and side c is hypotenuse.

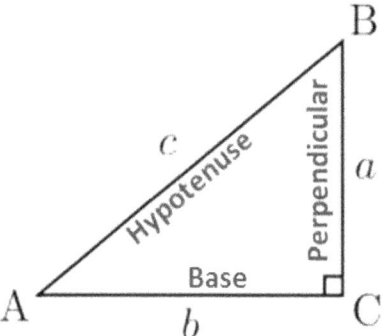

As mentioned earlier, Base is also called Adjacent (adj); and Perpendicular is also called Opposite (opp).

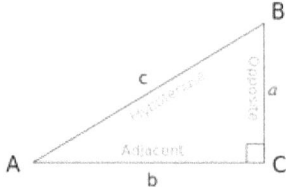

The ratios for given triangles below are as follows:

- Sine A = sin A = $\dfrac{\text{perpendicular}}{\text{hypotenuse}} = \dfrac{a}{c} = \dfrac{opp}{hyp}$

- Cosine A = cos A = $\dfrac{\text{base}}{\text{hypotenuse}} = \dfrac{b}{c} = \dfrac{adj}{hyp}$

- Tangent A = tan A = $\dfrac{\text{perpendicular}}{\text{base}} = \dfrac{a}{b} = \dfrac{opp}{adj}$

Trigger: **SOHCAHTOA** (SOH=>Sin=Opp/Hyp, CAH=>Cos=Adj/Hyp)

Or,

Trigger: **SPHCBHTPB** (SPH =>Sin=Perp/Hyp, CBH=>Cos=Base/Hyp)

And the three reciprocal identities are:

- Cosecant A = cosec A = $\dfrac{\text{hypotenuse}}{\text{perpendicular}} = \dfrac{c}{a} = \dfrac{hyp}{opp} = \dfrac{1}{\sin A}$

- Secant A = sec A = $\dfrac{\text{hypotenuse}}{\text{base}} = \dfrac{c}{b} = \dfrac{hyp}{adj} = \dfrac{1}{\cos A}$

- Cotangent A = cot A = $\dfrac{\text{base}}{\text{perpendicular}} = \dfrac{b}{a} = \dfrac{adj}{opp} = \dfrac{1}{\tan A}$

2.2.1 Values of Trigonometric Ratios

All the trigonometric ratios for angle of measures 0°, 30°, 45°, 60°, 90° are provided in the following table:

θ	$0°$	$30°$	$45°$	$60°$	$90°$
$\sin \theta$	0	$\dfrac{1}{2}$	$\dfrac{1}{\sqrt{2}}$	$\dfrac{\sqrt{3}}{2}$	1
$\cos \theta$	1	$\dfrac{\sqrt{3}}{2}$	$\dfrac{1}{\sqrt{2}}$	$\dfrac{1}{2}$	0
$\tan \theta$	0	$\dfrac{1}{\sqrt{3}}$	1	$\sqrt{3}$	Not defined
$\cot \theta$	Not defined	$\sqrt{3}$	1	$\dfrac{1}{\sqrt{3}}$	0
$\sec \theta$	1	$\dfrac{2}{\sqrt{3}}$	$\sqrt{2}$	2	Not defined
$\operatorname{cosec} \theta$	Not defined	2	$\sqrt{2}$	$\dfrac{2}{\sqrt{3}}$	1

Note: Observe the values of sin and cos as complementary to each other: sin 0° = cos 90°; sin 30° = cos 60°; sin 45° = cos 45°; sin 90°=cos 0°;

The trigonometric functions: sin and cos; cosec and sec; tan and cot; are complementary to each other.

2.3   Derived Trigonometric Ratios

Relationships among trigonometric functions are summarized below:

There are six trigonometric ratios and some well-defined notation to express the fundamental identities. Trigonometric identities are equalities that involve trigonometric functions and are true for every single value of the occurring variables.

If understood, memorize the following basic identities to solve the trigonometric puzzles:

Reciprocal Identities:

- $cosec\ A = \dfrac{1}{\sin A}$

- $cot\ A = \dfrac{1}{\tan A}$

- $sec\ A = \dfrac{1}{\cos A}$

Quotient Identities:

- $tan\ u = \dfrac{\sin u}{\cos u}$

- $cot\ u = \dfrac{\cos u}{\sin u}$

Reciprocal and Quotient Identities with Conditions:

As we know, sin and cos produce 0 for certain angles, so conditions are attached for basic trigonometric identities discussed earlier.

The denominator of any term should not be 0, otherwise the expression becomes undefined. The identities with conditions are as follows:

Reciprocal Identities with conditions:

- $cosec\ A = \dfrac{1}{\sin A}$, $A \neq n\pi$, where n is any integer

- $cot\ A = \dfrac{1}{\tan A}$, $A \neq n\pi$, where n is any integer

- $sec\ A = \dfrac{1}{\cos A}$, $A \neq (2n+1)\dfrac{\pi}{2}$, where n is any integer

---

Quotient Identities with conditions:

- $tan\ u = \dfrac{\sin u}{\cos u}$, $u \neq (2n+1)\dfrac{\pi}{2}$, where n is any integer

- $cot\ u = \dfrac{\cos u}{\sin u}$, $u \neq n\pi$, where n is any integer

---

Example-1: To find the height of a tree, a person walks to a point 20 feet from the base of the tree, and measures the angle from the ground to the top of the tree to be 30 degrees. Find the height of the tree.

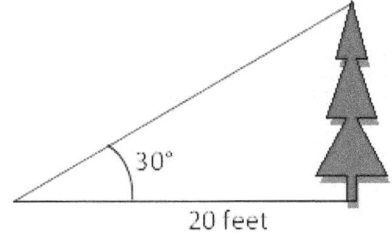

Solution: We know that, $\tan \beta = \dfrac{\text{perpendicular}}{\text{base}}$

Let the height of tree be h, so $\tan 30° = \dfrac{h}{20}$

As, $\tan 30° = \dfrac{1}{\sqrt{3}}$

Therefore, $h = \dfrac{20}{\sqrt{3}}$ feet

2.4   Polar Coordinate Trigonometry

A coordinate system specifies a point, or coordinate, to uniquely determine its position.

Cartesian Coordinate System

Cartesian coordinate system also called Rectangular coordinate system, uses the perpendicular lines, like X and Y-axis to specify the position of a point in a two-dimensional plane.

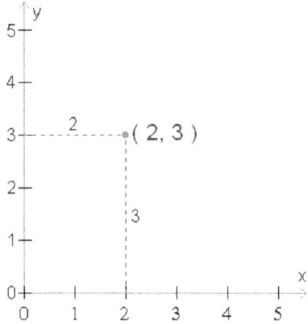

Cartesian coordinate is in the form (x, y), where 'x' and 'y' are the horizontal and vertical distances from the origin. A point (2, 3) is depicted in Cartesian coordinate system as shown in the figure above.

Polar Coordinate System

A polar coordinate system is a two-dimensional coordinate system in which each point on a plane is determined by a distance from the pole (analogous to origin of a Cartesian system) and an angle from the polar axis (analogous to positive X-axis of a Cartesian system).

Polar coordinate of a point is in the form (r, θ), where r is the radius (also called radial coordinate) from the pole (origin) to the point; and θ is polar angle (also called angular coordinate) measured from the polar axis (positive x-axis) to the radius.

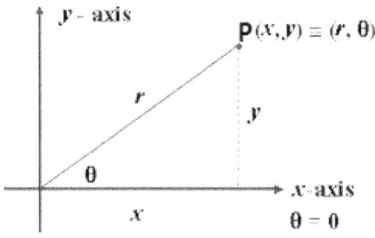

The polar coordinate of the point P denoted as (r, θ) in terms of (radius, angle) is depicted in the figure. This point P may also be represented by many pairs of polar coordinates, like $(r, \theta+2\pi)$ and $(-r, \theta+\pi)$.

In trigonometry, we can convert the position of a point from Cartesian coordinate to Polar coordinate and vice versa.

Converting Polar to Rectangular Coordinate:

To convert a polar coordinate into rectangular coordinate, we use the trigonometric ratios sin and cos.

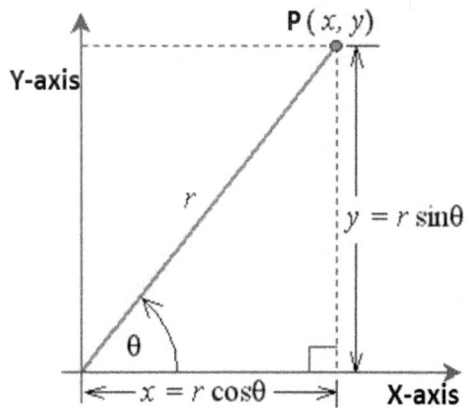

In the given figure, trigonometric ratio $cos\ \theta = \dfrac{x}{r}$, so

$$x = r\ cos\ \theta, \quad and \quad y = r\ sin\ \theta$$

Therefore, rectangular coordinate (r cos θ, r sin θ) is equivalent to polar coordinate (r, θ).

$(r, \theta) => (r\ cos\ \theta, r\ sin\ \theta)$

Example-1: A radius of 6 cm sweeps out an angle of 30° in standard position. What are the Cartesian coordinates (x, y) of the endpoint of the radius?

Solution: As given, the polar coordinate is (6, 30°). Transforming into Cartesian coordinate,

$$x = r\ cos\ \theta = 6\ cos\ 30° = 6 \times \frac{\sqrt{3}}{2} = 3\ \sqrt{3}$$

$$y = r\ sin\ \theta = 6\ sin\ 30° = 6 \times \frac{1}{2} = 3$$

So, the Cartesian coordinate is $(3\ \sqrt{3},\ 3)$

Converting Rectangular to Polar Coordinate:

To convert a polar coordinate into rectangular coordinate, we use the Pythagorean theorem and trigonometric ratio tan.

As per Pythagorean theorem,

$$r = \sqrt{x^2 + y^2}$$

And $tan\ \theta = \frac{y}{x}$, So angle θ is written in form of inverse tan and denoted as tan^{-1},

$$\theta = tan^{-1} \frac{y}{x}$$

Therefore, polar coordinate $(\sqrt{x^2 + y^2},\ tan^{-1} \frac{y}{x})$ is equivalent of rectangular coordinate (x, y).

$$(x, y) => (\sqrt{x^2 + y^2},\ tan^{-1} \frac{y}{x})$$

Example-2: A point is specified as rectangular coordinate (3, 4). Find the position of this point in polar coordinate.

Solution: As given, the rectangular coordinate is (3, 4). Transforming into polar coordinate,

$$r = \sqrt{x^2 + y^2} = \sqrt{3^2 + 4^2} = 5$$

$$\theta = tan^{-1} \frac{y}{x} = tan^{-1} \frac{4}{3} = tan^{-1} 1.334 = 53° \text{ (Approx.)}$$

So, the polar coordinate is (5, *53°*)

(*Note: refer the appendix to find angle for value of tan as 1.334*)

Worksheet-5

Exercise-1: Based on reciprocal and quotient identities, verify the following:

Exercise-1-a: $\cot x \tan x = 1$

Exercise-1-b: $\csc x \, \tan x = \sec x$

Exercise-1-c: $\csc x \sin x = 1$

Exercise-1-d: $\sec x \cot x = \csc x$

Exercise-1-e: $\sec x \csc x = \dfrac{1}{\cos x \sin x}$

Exercise-1-f: $\dfrac{1 + \sin x}{\cos x} = \sec x + \tan x$

Exercise-2: Fill the values of trigonometric functions for different angles in the table.

β=	0°	30°	45°	60°	90°
sin β					
cos β					
tan β					
cot β					
sec β					
cosec β					

Exercise-3: A person is standing on the level ground, at B, a distance of 19 meters away from the foot E of a tree TE. He measures the angle from the ground to the top of the tree to be 30 degrees. Find the height of the tree.

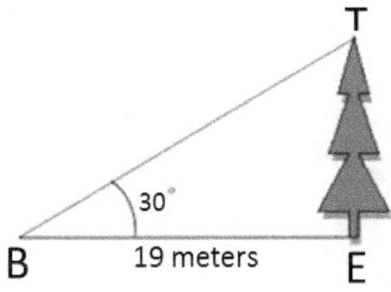

B   19 meters   E

Exercise-4: A person at a point 10 feet from the base of the tree measures the angle from the ground to the top of the tree to be 45 degrees. Find the height of the tree.

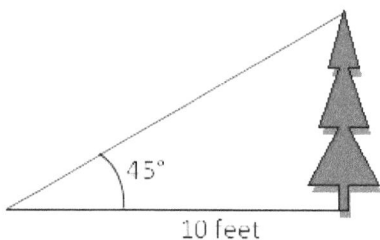

Exercise-5: A 25-feet ladder leans against a wall so that the angle between the ground and the ladder is 60°. How high does the ladder reach up the side of the wall?

Exercise-6: In the following figure, find the length x

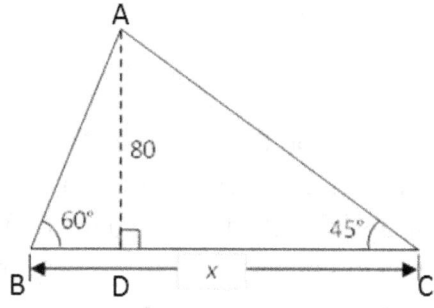

Exercise-7: In the following figure, find the length y.

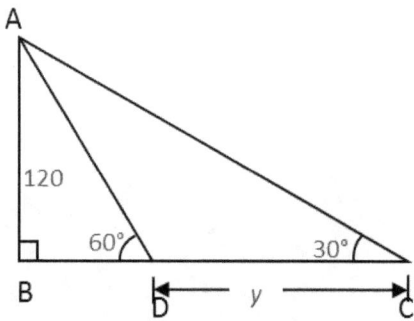

Exercise-8: A person on the roof of a 200 foot building at point F looks towards a skyscraper AB wishes to measure its height. He measures the angle of declination from the roof of the building F to the base of the skyscraper B to be 30 degrees and the angle of inclination to the top of the skyscraper A to be 45 degrees. Find the height of skyscraper AB.

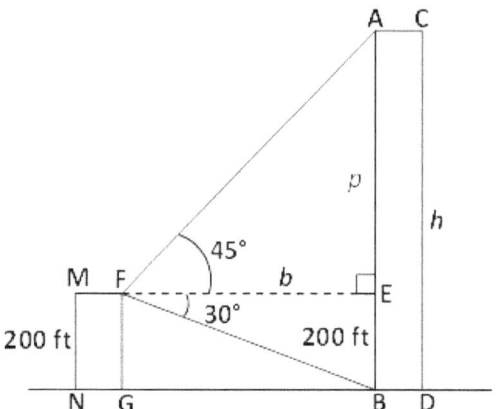

Exercise-9: Convert a point (10, 30°) on polar coordinate system to rectangular coordinate.

Exercise-10: Convert a point (5, 12) on rectangular coordinate system to polar coordinate.

Solutions-Worksheet-5

Solution-1-a: Converting the cot into tan, we have

$$\frac{1}{\tan x} \text{ x } \tan x = 1 \text{ } \underline{Q.E.D}$$

Solution-1-b: Converting the cosec into sin; and tan into sin and cos, we have

$$\frac{1}{\sin x} \text{ x } \frac{\sin x}{\cos x} = \frac{1}{\cos x} = \sec x \qquad \underline{Q.E.D}$$

Solution-1-c: Converting the cosec into sin, we have

$$\frac{1}{\sin x} \text{ x } \sin x = 1 \qquad \underline{Q.E.D}$$

Solution-1-d: Converting the sec into cos; and cot into sin and cos, we have

$$\frac{1}{\cos x} \text{ x } \frac{\cos x}{\sin x} = \frac{1}{\sin x} = \text{cosec } x \qquad \underline{Q.E.D}$$

Solution-1-e: Converting the sec into cos; and cosec into sin, we have

$$\frac{1}{\cos x} \text{ x } \frac{1}{\sin x} = \frac{1}{\cos x \, \sin x} \qquad \underline{Q.E.D}$$

Solution-1-f: Separating the left hand side (LHS) pf equation into two terms and then changing into sec and tan, we have

$$\frac{1 + \sin x}{\cos x} = \frac{1}{\cos x} + \frac{\sin x}{\cos x} = \sec x + \tan x \qquad \underline{Q.E.D}$$

Solution-2:

θ	$0°$	$30°$	$45°$	$60°$	$90°$
$\sin \theta$	0	$\dfrac{1}{2}$	$\dfrac{1}{\sqrt{2}}$	$\dfrac{\sqrt{3}}{2}$	1
$\cos \theta$	1	$\dfrac{\sqrt{3}}{2}$	$\dfrac{1}{\sqrt{2}}$	$\dfrac{1}{2}$	0
$\tan \theta$	0	$\dfrac{1}{\sqrt{3}}$	1	$\sqrt{3}$	Not defined
$\cot \theta$	Not defined	$\sqrt{3}$	1	$\dfrac{1}{\sqrt{3}}$	0
$\sec \theta$	1	$\dfrac{2}{\sqrt{3}}$	$\sqrt{2}$	2	Not defined
$\csc \theta$	Not defined	2	$\sqrt{2}$	$\dfrac{2}{\sqrt{3}}$	1

Solution-3: We need to find the perpendicular, while base and angle are given. Tangent relates the perpendicular with base. Let's assume the height of tree is h. Therefore,

$$\tan 30° = \frac{h}{19}$$

$$h = 19 \left(\frac{1}{\sqrt{3}}\right) = 10.97 \text{ meters (approx.)}$$

[*Refer table of Appendix-1: tan 30° = 0.5774 approx.*]

Solution-4: We need to find the perpendicular, while base and angle are given. Tangent relates the perpendicular with base. Let's assume the height of tree is h. Therefore,

$$\tan 45° = \frac{h}{10}$$

$$h = 10 \times 1 = 10 \text{ feet}$$

Solution-5: We need to find the perpendicular, while hypotenuse and angle are given. Sine relates the perpendicular with hypotenuse. Let's assume the height of wall is p. Therefore,

$$\sin 60° = \frac{p}{25}$$

$$p = 25 \times \frac{\sqrt{3}}{2} = 21.65 \text{ feet}$$

Solution-6: We will deal only with right triangle to apply our trigonometric functions. Therefore, to find the base x, we will find the base BD and base DC and then add them up. Tangent is the ratio of perpendicular and base, so

$$BD = \frac{80}{\tan 60°} = \frac{80}{1.732} = 46.19 \text{ (approx.)}$$

$$DC = \frac{80}{\tan 45°} = \frac{80}{1} = 80$$

$x = BD + DC = 46.19 + 80 = 126.19$ (approx.)

Solution-7: We will deal only with right triangle to apply our trigonometric functions. Therefore, to find y, we will find BC and BD, thereafter subtract BD from BC.

$$BC = \frac{120}{\tan 30°} = \frac{120}{0.5774} = 207.83 \text{ (approx.)}$$

$$BD = \frac{120}{\tan 60°} = \frac{120}{1.732} = 69.28 \text{ (approx.)}$$

$y = BC - BD = 207.97 - 69.28 = 138.69$ (approx.)

(Refer Appendix-1: Value of tan 30° = 0.5774; Value of tan 60° = 1.732)

Solution-8: We have to deal with two right triangles AFE and EFB. We will consider perpendicular h (AE+EB) and common base b for the given angles.

In right triangle FEA, we have: AE = b (tan 45°) = b   ---Eq(1)

In right triangle FEB, we have: $b = \dfrac{200}{\tan 30°} = 346.38$ ---Eq(2)

(Refer Appendix-1: Value of tan 30° = 0.5774)

So substituting the value of b from Eq(2) in Eq(1), we get

     AE = 346.38 feet

Therefore, h = AE + EB = 346.38 + 200 = 546.38 feet

Solution-9: $(x, y) = (10 \cos 30°, 10 \sin 30°) = (5\sqrt{3}, 5)$

Solution-10: $r = \sqrt{5^2 + 12^2} = 13$; $\theta = \tan^{-1}(12/5) = 68°$ (approx.) so $(r, \theta) = (13, 68°)$

Chapter 3: Trigonometric Functions, Quadrantal Values, and Graphs

3.1 Pythagorean Identities

We will now study the trigonometric ratios used in trigonometric identities and formulas. Consider a unit circle (radius is 1 unit) with centre at origin of the coordinate axis.

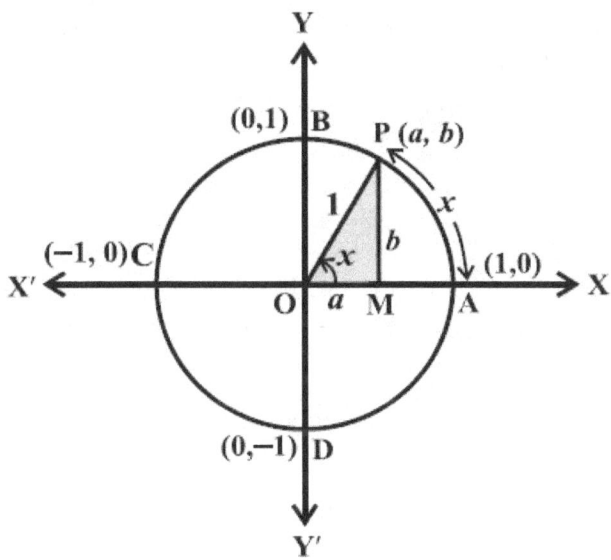

A unit circle containing a right triangle

The unit circle touches the coordinate axis at point A(1,0), B(0,1), C(−1,0) and D(0, −1). Let P (a, b) be any point on this unit circle where angle AOP = x radian, which is equal to length of arc AP.

$$\text{Angle AOP in radian} = \frac{\text{length of arc}}{\text{radius}} = \frac{x}{1} = x \text{ radian}$$

Proof of Pythagorean Identity:

Since OMP is a right triangle, from Pythagorean Theorem, we get

$$OM^2 + MP^2 = OP^2, \text{i.e. } a^2 + b^2 = 1$$

As per ratio definition of cos and sin, cos $x = a / 1 = a$; and sin $x = b / 1 = b$; Therefore, replacing a with sin x, and b with cos x, we get:

$$sin^2x + cos^2x = 1$$

Note: This is an identity categorized under Pythagorean Identities. This identity holds true for Point P (a, b) to be anywhere in the circle. So this equation is valid for all real x.

<u>Proof using Right Triangle</u>: Let's investigate the identity directly using Pythagorean Theorem. As shown in figure below, ABC is a right triangle with sides a, b, and c with right angle at C.

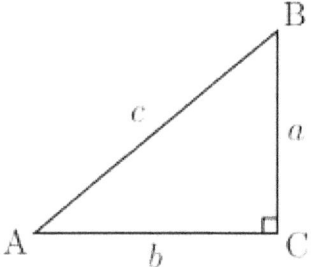

We know the ratios of sides as: $\sin A = \dfrac{a}{c}$ and $\cos A = \dfrac{b}{c}$

Therefore,

$$\sin^2 A + \cos^2 A = (\frac{a}{c})^2 + (\frac{b}{c})^2$$

$$= (\frac{a}{c})^2 + (\frac{b}{c})^2 = \frac{a^2}{c^2} + \frac{b^2}{c^2} = \frac{a^2 + b^2}{c^2}$$

Using Pythagorean Theorem $a^2 + b^2 = c^2$,

$$\Rightarrow \frac{a^2 + b^2}{c^2} = \frac{c^2}{c^2} = 1$$

Hence, $\sin^2 A + \cos^2 A = 1$

This is the first Pythagorean identity.

Other two Pythagorean identities

If the Pythagorean identity $sin^2\ A + cos^2\ A = 1$ is divided by $\cos^2 A$,

$$\frac{\sin^2 A}{\cos^2 A} + \frac{\cos^2 A}{\cos^2 A} = \frac{1}{\cos^2 A}$$

We get another identity as: $\tan^2 A + 1 = \sec^2 A$

Similarly, if the Pythagorean identity $sin^2\ A + cos^2\ A = 1$ is divided by $\sin^2 A$,

$$\frac{\sin^2 A}{\sin^2 A} + \frac{\cos^2 A}{\sin^2 A} = \frac{1}{\sin^2 A}$$

We get yet another identity as: $1 + \cot^2 A = cosec^2 A$

Pythagorean Identities:

- $\sin^2 \alpha + \cos^2 \alpha = 1$
- $1 + \tan^2 \alpha = \sec^2 \alpha$
- $1 + \cot^2 \alpha = \mathrm{cosec}^2 \alpha$

Using Pythagorean identities, trigonometric functions can be transformed in terms of others as shown below:

in terms of	$\sin\theta$	$\cos\theta$	$\tan\theta$
$\sin\theta =$	$\sin\theta$	$\pm\sqrt{1-\cos^2\theta}$	$\pm\dfrac{\tan\theta}{\sqrt{1+\tan^2\theta}}$
$\cos\theta =$	$\pm\sqrt{1-\sin^2\theta}$	$\cos\theta$	$\pm\dfrac{1}{\sqrt{1+\tan^2\theta}}$
$\tan\theta =$	$\pm\dfrac{\sin\theta}{\sqrt{1-\sin^2\theta}}$	$\pm\dfrac{\sqrt{1-\cos^2\theta}}{\cos\theta}$	$\tan\theta$
$\csc\theta =$	$\dfrac{1}{\sin\theta}$	$\pm\dfrac{1}{\sqrt{1-\cos^2\theta}}$	$\pm\dfrac{\sqrt{1+\tan^2\theta}}{\tan\theta}$
$\sec\theta =$	$\pm\dfrac{1}{\sqrt{1-\sin^2\theta}}$	$\dfrac{1}{\cos\theta}$	$\pm\sqrt{1+\tan^2\theta}$
$\cot\theta =$	$\pm\dfrac{\sqrt{1-\sin^2\theta}}{\sin\theta}$	$\pm\dfrac{\cos\theta}{\sqrt{1-\cos^2\theta}}$	$\dfrac{1}{\tan\theta}$

Worksheet-6

Exercise-1: Verify the following:

1-a: $\sin x \tan x + \cos x = \sec x$

1-b: $\cos x \cot x + \sin x = \operatorname{cosec} x$

1-c: $\sin x - \sin x \cos^2 x = \sin^3 x$

1-d: $\cos x - \cos x \sin^2 x = \cos^3 x$

1-e: $\dfrac{\cos x}{1 + \sin x} + \dfrac{1 + \sin x}{\cos x} = 2 \sec x$

1-f: $\dfrac{\sin x}{1 + \cos x} = \dfrac{1 - \cos x}{\sin x}$

Exercise-2: Transform the trigonometric function in terms of other as required in following table.

	sin α	cos α	tan α
sin α =			
cos α =			
tan α =			
cosec α =			
sec α =			
cot α =			

Exercise-3: Prove the identity:

$$1 + \tan^2 u = \sec^2 u$$

Exercise-4: Prove the identity:

$$1 + \cot^2 u = \csc^2 u$$

Exercise-5: Verify

$$\frac{\sec(t) - \cos(t)}{\sin(t)} = \tan(t)$$

Exercise-6: Verify

$$\frac{1 + \cot(t)}{1 + \tan(t)} = \cot(t)$$

Exercise-7: Verify

$$\frac{\sin^2(t) + \cos^2(t)}{\cos^2(t)} = \sec^2(t)$$

Exercise-8: Prove that

$$\frac{\sin^2(\theta)}{1 + \cos(\theta)} = 1 - \cos(\theta)$$

Exercise-9: Prove that

$$\sec(a) - \cos(a) = \sin(a)\tan(a)$$

Exercise-10: Verify

$$\frac{\cosec^2(x) - \sin^2(x)}{\cosec(x) + \sin(x)} = \cos(x)\cot(x)$$

Exercise-11: Verify

$$\frac{\cosec^2(\alpha) - 1}{\cosec^2(\alpha) - \cosec(\alpha)} = 1 + \sin(\alpha)$$

Exercise-12: Verify

$$\frac{1+\cos(u)}{\sin(u)} = \frac{\sin(u)}{1-\cos(u)}$$

Exercise-13: Verify

$$\frac{\sin^4(\gamma)-\cos^4(\gamma)}{\sin(\gamma)-\cos(\gamma)} = \sin(\gamma)+\cos(\gamma)$$

Solutions-Worksheet-6

Solution-1-a: changing tan into sin and cos, the left hand side (LHS) of equation is,

$$\sin x \; \frac{\sin x}{\cos x} + \cos x$$

$$= \frac{\sin^2 x}{\cos x} + \cos x = \frac{\sin^2 x + \cos^2 x}{\cos x} = \frac{1}{\cos x} = \sec x \quad \underline{Q.E.D.}$$

Solution-1-b: changing cot into sin and cos, the left hand side (LHS) of equation is,

$$\cos x \; \frac{\cos x}{\sin x} + \sin x$$

$$= \frac{\cos^2 x}{\sin x} + \sin x = \frac{\cos^2 x + \sin^2 x}{\sin x} = \frac{1}{\sin x} = \mathrm{cosec}\, x \quad \underline{Q.E.D.}$$

Solution-1-c: Taking sin as common and replacing $(1 - \cos^2 x) = \sin^2 x$ in LHS of equation, we have

$$\sin x \,(1 - \cos^2 x) = \sin x \,(\sin^2 x) = \sin^3 x \qquad \underline{Q.E.D}$$

Solution-1-d: Taking cos as common and replacing $(1 - \sin^2 x) = \cos^2 x$ in LHS of equation, we have

$$\cos x \,(1 - \sin^2 x) = \cos x \,(\cos^2 x) = \cos^3 x \qquad \underline{Q.E.D}$$

Solution-1-e: Simplifying the LHS of equation,

$$\frac{\cos^2 x + (1+\sin x)^2}{(1 + \sin x)(\cos x)} = \frac{\cos^2 x + (1+\sin^2 x + 2\sin x)}{(1 + \sin x)(\cos x)}$$

$$= \frac{\cos^2 x + \sin^2 x + 1 + 2\sin x)}{(1 + \sin x)(\cos x)} = \frac{2 + 2\sin x}{(1 + \sin x)(\cos x)} =$$

$$= \frac{2(1+\sin x)}{(1 + \sin x)(\cos x)} = \frac{2}{\cos x} = 2 \sec x \qquad \text{Q.E.D}$$

Solution-1-f: Moving the terms of RHS at LHS of equation, we have,

$$\frac{\sin x}{(1 + \cos x)} \times \frac{\sin x}{(1 - \cos x)} = 1$$

Simplifying the LHS of this equation,

$$\frac{\sin^2 x}{(1 - \cos^2 x)} = \frac{\sin^2 x}{\sin^2 x} = 1 \qquad \text{Q.E.D.}$$

Solution-2: Refer for table of transformation in the section 3.1 of this chapter.

Solution-3: To prove the identity, we replace $tan^2\ u$ with $\dfrac{sin^2 u}{cos^2 u}$

and use the theorem, $sin^2\ u + cos^2\ u = 1$

$$1 + \frac{sin^2 u}{cos^2 u} = \frac{cos^2 u + sin^2 u}{cos^2 u} = \frac{1}{cos^2 u} = \sec^2 u \qquad \text{Q.E.D}$$

Solution-4: To prove the identity, we replace $cot^2\ u$ with $\dfrac{cos^2 u}{sin^2 u}$

and use the theorem $\cos^2 u + \sin^2 u = 1$

$$\frac{\cos^2 u + \sin^2 u}{\sin^2 u} = \frac{1}{\sin^2 u}$$

$$= \operatorname{cosec}^2 u \qquad \text{Q.E.D}$$

Solution-5:

$$\frac{\sec(t)-\cos(t)}{\sin(t)} = \frac{\frac{1}{\cos(t)}-\cos(t)}{\sin(t)} = \frac{1-\cos^2(t)}{\sin(t)\cos(t)} = \frac{\sin^2(t)}{\sin(t)\cos(t)} =$$

$$\frac{\sin(t)}{\cos(t)} = \tan(t) \qquad \text{Q.E.D}$$

Solution-6:

$$\frac{1+\cot(t)}{1+\tan(t)} = \frac{1+\frac{1}{\tan(t)}}{1+\tan(t)} \cdot \frac{\tan(t)}{\tan(t)} = \frac{\tan(t)+1}{(1+\tan(t))\tan(t)} = \frac{1}{\tan(t)}$$

$$= \cot(t) \qquad \text{Q.E.D}$$

Solution-7:

$$\frac{\sin^2(t)+\cos^2(t)}{\cos^2(t)} = \frac{1}{\cos^2(t)} = \sec^2(t) \qquad \text{Q.E.D}$$

Solution-8: Working at LHS of equation, we apply the Pythagorean identity $\sin^2(\theta) + \cos^2(\theta) = 1$

$$\frac{\sin^2(\theta)}{1+\cos(\theta)} = \frac{1-\cos^2(\theta)}{1+\cos(\theta)}$$

$$= \frac{(1+\cos(\theta))(1-\cos(\theta))}{1+\cos(\theta)}$$

$$= 1 - \cos(\theta) \qquad \text{Q.E.D}$$

Solution-9: Working at LHS of equation, we have,

$$\sec(a) - \cos(a) = \frac{1}{\cos(a)} - \cos(a)$$

$$= \frac{1}{\cos(a)} - \frac{\cos^2(a)}{\cos(a)}$$

$$= \frac{\sin^2(a)}{\cos(a)}$$

$$= \sin(a) \cdot \frac{\sin(a)}{\cos(a)}$$

$$= \sin(a) \cdot \tan(a)$$

Solution-10: Working at LHS of equation, we factorize the expression,

$$\frac{\cosec^2(x) - \sin^2(x)}{\cosec(x) + \sin(x)} = \frac{(\cosec(x) + \sin(x))(\cosec(x) - \sin(x))}{\cosec(x) + \sin(x)}$$

$$= \cosec(x) - \sin(x)$$

$$= \frac{1}{\sin(x)} - \sin(x)$$

$$= \frac{1}{\sin(x)} - \frac{\sin^2(x)}{\sin(x)}$$

$$= \frac{\cos^2(x)}{\sin(x)}$$

$$= \cos(x) \cdot \frac{\cos(x)}{\sin(x)}$$

$$= \cos(x) \cot(x) \qquad \text{Q.E.D}$$

Solution-11: Working at LHS of equation, we have

$$\frac{\cosec^2(\alpha) - 1}{\cosec^2(\alpha) - \cosec(\alpha)} = \frac{(\cosec(\alpha) + 1)(\cosec(\alpha) - 1)}{\cosec(\alpha) \cdot (\cosec(\alpha) - 1)}$$

$$= \frac{\cosec(\alpha)+1}{\cosec(\alpha)} = \frac{\frac{1}{\sin(\alpha)}+1}{\frac{1}{\sin(\alpha)}}$$

$$= \left(\frac{1}{\sin(\alpha)}+1\right) \cdot \frac{\sin(\alpha)}{1} = \frac{\sin(\alpha)}{\sin(\alpha)} + \frac{\sin(\alpha)}{1}$$

$$= 1 + \sin(\alpha) \qquad \text{Q.E.D}$$

Solution-12: At the RHS of equation, $\sin(u)$ is the numerator. To get that, we multiply and divide the LHS by $1 - \cos(u)$ and use the Pythagorean identity:

$$\frac{1+\cos(u)}{\sin(u)} \cdot \frac{1-\cos(u)}{1-\cos(u)} = \frac{1-\cos^2(u)}{\sin(u)(1-\cos(u))}$$

$$= \frac{\sin^2(u)}{\sin(u)(1-\cos(u))}$$

$$= \frac{\sin(u)}{1-\cos(u)}$$

Solution-13: Working at LHS of equation, we factorize the expression,

$$\frac{\sin^4(\gamma)-\cos^4(\gamma)}{\sin(\gamma)-\cos(\gamma)} = \frac{(\sin^2(\gamma)+\cos^2(\gamma))\,(\sin^2(\gamma)-\cos^2(\gamma))}{\sin(\gamma)-\cos(\gamma)}$$

$$= \frac{1\cdot(\sin(\gamma)+\cos(\gamma))(\sin(\gamma)-\cos(\gamma))}{\sin(\gamma)-\cos(\gamma)}$$

$$= \sin(\gamma) + \cos(\gamma) \qquad \text{Q.E.D}$$

3.2   Trigonometric Functions in different Quadrants

X and Y axis of Cartesian-Coordinate System divide the plane into 4 different quadrants. Quadrants are numbered as I, II, III, IV in a counter-clockwise direction.

- In Quadrant I, both x and y are positive
- In Quadrant II, x is negative but y is positive
- In Quadrant III, both x and y are negative
- In Quadrant IV, x is positive but y is negative.

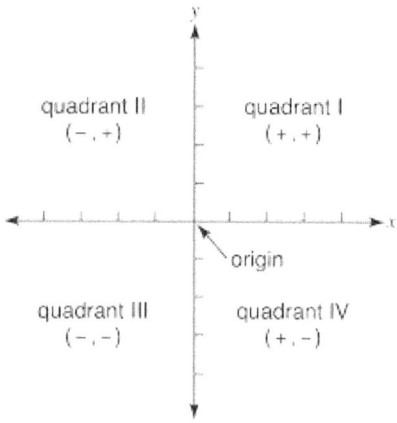

3.2.1   Quadrantal Angles

We refer to figure below of a unit circle containing a right triangle to explore the value of trigonometric functions in quadrantal angles. Some of the quadrantal angles are $\angle AOB = \pi/2$, $\angle AOC = \pi$ and $\angle AOD = 3\pi/2$. All angles which are multiples of $\pi/2$ are called quadrantal angles. The coordinates of the points are A(1,0), B(0,1), C(–1,0) and D(0,–1).

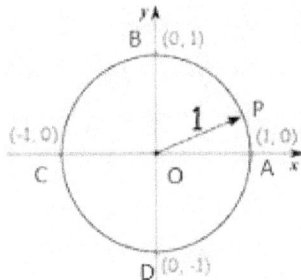

Values of Quadrantal Angles: For quadrantal angles, referring the figure above, we get the value:

$cos\ 0° = 1,\quad sin\ 0° = 0;$

$cos\ \dfrac{\pi}{2} = 0,\quad sin\ \dfrac{\pi}{2} = 1;$

$cos\ \pi = –1,\quad sin\ \pi = 0;$

$cos\ \dfrac{3\pi}{2} = 0,\quad sin\ \dfrac{3\pi}{2} = –1;$

$cos\ 2\pi = 1,\quad sin\ 2\pi = 0;$

Values of trigonometric ratios for some important angles are summarized in table which should be understood and memorized.

	$0°$	$\dfrac{\pi}{6}$	$\dfrac{\pi}{4}$	$\dfrac{\pi}{3}$	$\dfrac{\pi}{2}$	π	$\dfrac{3\pi}{2}$	2π
sin	0	$\dfrac{1}{2}$	$\dfrac{1}{\sqrt{2}}$	$\dfrac{\sqrt{3}}{2}$	1	0	-1	0
cos	1	$\dfrac{\sqrt{3}}{2}$	$\dfrac{1}{\sqrt{2}}$	$\dfrac{1}{2}$	0	-1	0	1
tan	0	$\dfrac{1}{\sqrt{3}}$	1	$\sqrt{3}$	not defined	0	not defined	0

Complete revolutions $2n\pi$: If one or several complete revolutions are traversed from a specific point, we come back to a same point. Therefore,

$sin\ (2n\pi + x) = sin\ x,\quad where\ n \in \mathbf{Z}$ (set of integers)

$cos\ (2n\pi + x) = cos\ x,\quad where\ n \in \mathbf{Z}$ (set of integers)

$tan\ (2n\pi + x) = tan\ x,\quad where\ n \in \mathbf{Z}$ (set of integers)

Even if we make revolution backward in clockwise direction where angle is negative, we get:

$sin\ (x - 2n\pi) = sin\ x,\quad where\ n \in \mathbf{Z}$ (set of integers)

$cos\ (x - 2n\pi) = cos\ x,\quad where\ n \in \mathbf{Z}$ (set of integers)

$tan\ (x - 2n\pi) = tan\ x,\quad where\ n \in \mathbf{Z}$ (set of integers)

Periodicity of functions:

$sin\ (x \pm 2n\pi) = sin\ x,$ **where** $n \in \mathbf{Z}$ **(set of integers)**

$cos\ (x \pm 2n\pi) = cos\ x,$ **where** $n \in \mathbf{Z}$ **(set of integers)**

$tan\ (x \pm 2n\pi) = tan\ x,$ **where** $n \in \mathbf{Z}$ **(set of integers)**

Note: sin and cos are periodic with 2π; tan is periodic with π, hence also with 2π.

Example-1: Find the value of $\sin 390°$

Solution: sin is periodic with 2π, so

$$\sin 390° = \sin (360° + 30°) = \sin 30° = \frac{1}{2}$$

Alternatively,

$$\sin 390° = \sin (390° - 360°) = \sin 30° = \frac{1}{2}$$

Example-2: Find the value of $\cos 750°$

Solution: cos is periodic with 2π, so

$$\cos 750° = \cos (2 \times 360° + 30°) = \cos 30° = \frac{\sqrt{3}}{2}$$

Alternatively,

$$\cos 750° = \cos (750° - 2 \times 360°) = \cos 30° = \frac{\sqrt{3}}{2}$$

Example-3: Find the value of $\cos -1750°$

Solution: cos is periodic with 2π, so

$$\cos (-1710°) = \cos (-1710° + 5 \times 360°)$$

$$= \cos (-1710° + 1800°) = \cos 90° = 0$$

Alternatively,

$$\cos (-1710°) = \cos (5 \times (-360°) + 90°) = \cos 90° = 0$$

Solution for sin/cos equals 0:

We have observed the quadrantal angles where the value of sin and cos are 0.

sin x = 0, implies x=0, ± π, ± 2π, ± 3π, ...

    So, x is an integral multiple of π

Therefore, general solution for *sin x=0* is

    $x = \pm n\pi$, where n ∈ **Z** (set of integers)

Similarly, cos x = 0, implies x= ± π/2, ± 3π/2, ± 5π/2, ...

    Here x is an odd integral multiple of π/2

Therefore, general solution for *cos x=0* is

    $x = \pm(2n+1)\dfrac{\pi}{2}$, where n ∈ **Z** (set of integers)

General Solution for sin, cos, tan be 0 :

sin x = 0, implies x = ±nπ , where n∈Z (set of integers)

cos x = 0, implies x= ±(2n+1)$\dfrac{\pi}{2}$,where n∈Z (set of integers)

tan x = 0, implies x = ±nπ , where n∈Z (set of integers)

3.2.2   Signs of Trigonometric Functions in different Quadrants

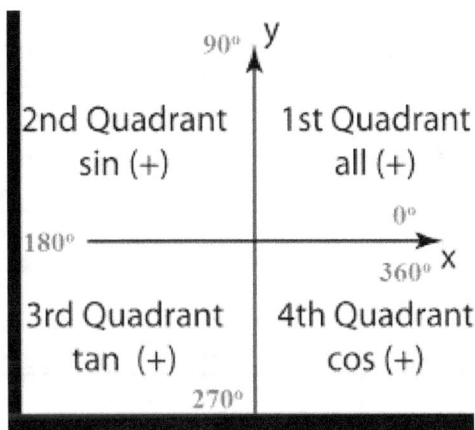

Signs of trigonometric functions in different quadrants:

- *sin, cos, tan, cosec, sec, cot; all   are positive in first quadrant*
- *cos and sec are negative in second and third quadrants*
- *sin and cosec are negative in third and fourth quadrant*
- *tan and cot are negative in second and fourth quadrants*

We will just understand and memorize the trigonometric functions sin, cos and tan because their corresponding reciprocal identities cosec, sec and cot follow the same sign in different quadrants.

- sin, cos, tan are positive in first quadrant; i.e. All (Trigger-A) are positive.
- cos and tan are negative in second quadrant; i.e. sin (Trigger-S) is positive

- sin and cos are negative in third quadrant; i.e. tan (Trigger-T) is positive
- sin and tan are negative in fourth quadrant; i.e. cos (Trigger-C) is positive

So, the trigger ASTC indicates positive sign of trigonometric functions in different quadrants.

Trigger: ASTC - All (Q-I), Sin(Q-II), Tan(Q-III), Cos(Q-IV)

Illustration of trigonometric functions in different quadrants

In Quadrant I, Sine, Cosine and Tangent are all positive. The Sine, Cosine and Tangent of 30° are:

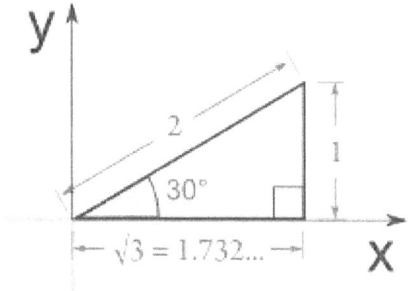

- sin(30°) = 1 / 2 = 0.5
- cos(30°) = 1.732 / 2 = 0.866
- tan(30°) = 1 / 1.732 = 0.577

In Quadrant II, as x is negative, both Cosine and Tangent become negative. The Sine, Cosine and Tangent of 150°:

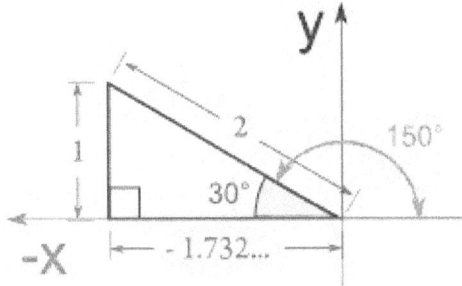

- $\sin(150°) = 1 / 2 = 0.5$
- $\cos(150°) = -1.732 / 2 = -0.866$
- $\tan(150°) = 1 / -1.732 = -0.577$

In Quadrant III, Sine and Cosine are negative. The Sine, Cosine and Tangent of 210° are:

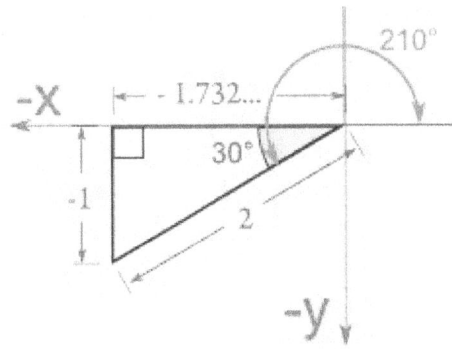

- $\sin(210°) = -1 / 2 = -0.5$
- $\cos(210°) = -1.732 / 2 = -0.866$
- $\tan(210°) = -1 / -1.732 = 0.577$

In Quadrant IV, Sine and Tangent are negative. The Sine, Cosine and Tangent of 330° are:

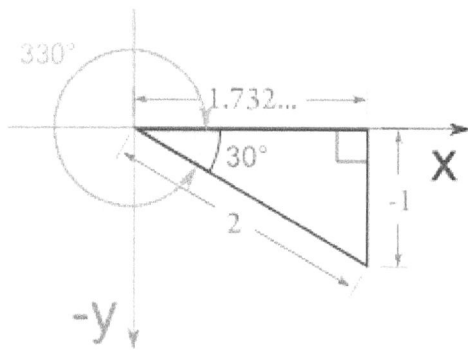

- $\sin(330°) = -1 / 2 = -0.5$
- $\cos(330°) = 1.732 / 2 = 0.866$
- $\tan(330°) = -1 / 1.732 = -0.577$

Remarks: In any quadrant, the sign of cosec is same as sign of sin, the sign of sec is same as sign of cos; the sign of tan depends upon sign of sin and cos; and the sign of cot is same as sign of tan.

Sign of sin, cos and tan functions in four quadrants:

- *All are positive in Q I.*
- *cos and sec are negative in Q II and Q III*
- *sin and cosec are negative in Q III & Q IV*
- *tan and cot are negative in Q II and Q IV*

Trigger: ***ASTC*** *- All (Q-I), Sin(Q-II), Tan(Q-III), Cos(Q-IV)*

Summary table of sign of trigonometric functions in different quadrants:

	I	II	III	IV
$\sin x$	+	+	−	−
$\cos x$	+	−	−	+
$\tan x$	+	−	+	−
$\operatorname{cosec} x$	+	+	−	−
$\sec x$	+	−	−	+
$\cot x$	+	−	+	−

Worksheet-7

Exercise-1: Find the sign of sin 390°

Exercise-2: Find the sign of cos 750°

Exercise-3: Find the sign of cos −1750°

Exercise-4: Write the general solution for sin x=0 and cos x=0

Exercise-5: Write values for the following trigonometric functions of quadrantal angles:

$cos\ 0° =$ _____ ; $sin\ 0° =$ _____ ;

$cos\ \dfrac{\pi}{2} =$ _____ ; $sin\ \dfrac{\pi}{2} =$ _____ ;

$cos\ \pi =$ _____ ; $sin\ \pi =$ _____ ;

Exercise-6: Write values for the following trigonometric functions of quadrantal angles:

$cos\ \dfrac{3\pi}{2} =$ _____ ; $sin\ \dfrac{3\pi}{2} =$ _____ ;

$cos\ 2\pi =$ _____ ; $sin\ 2\pi =$ _____ ;

Exercise-7: What are the signs of trigonometric functions in different quadrants? Write the trigger.

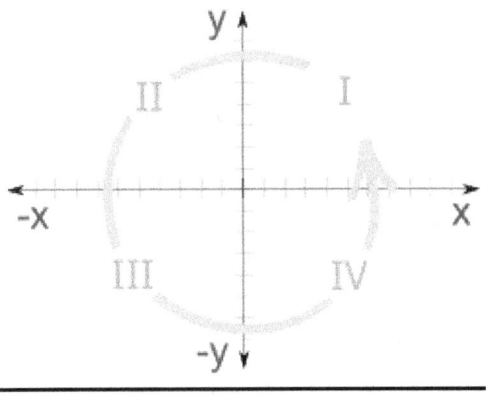

Exercise-8: Based on different quadrant, mention the appropriate sign (positive or negative) for following:

- sin(30°)
- cos(30°)
- tan(30°)
- sin(150°)
- cos(150°)
- tan(150°)
- sin(210°)
- cos(210°)
- tan(210°)
- sin(330°)
- cos(330°)
- tan(330°)

Solutions-Worksheet-7

Solution-1: If one or several complete revolutions are taken in positive (anticlockwise) or negative (clockwise) direction, we come back to the same point. This is termed as coterminal angle. So, *sin (x ± 2nπ) = sin x,   where* n ∈ **Z** (set of integers)

$390° = 360° + 30°$; Angle 30° is in first quadrant, so Sine of 390° is positive.

Solution-2 If one or several complete revolutions are taken in positive (anticlockwise) or negative (clockwise) direction, we come back to the same point. This is termed as coterminal angle. So, *cos (x ± 2nπ) = cos x, where* n ∈ **Z** (set of integers)

$750° = 2 \times 360° + 30°$; Angle 30° is in first quadrant, so Cosine of 750° is positive.

Solution-3: *cos (x ± 2nπ) = cos x, where* n ∈ **Z** (set of integers)

$-1750° = 5 \times (-360°) + 50°$; Angle 50° is in first quadrant, so Cosine of –1750° is positive.

 Solution-4: General solution for sin $x=0$ and cos $x=0$

sin $x = 0$, implies $x = \pm n\pi$, where n ∈ **Z** (set of integers)
cos $x = 0$, implies $x = \pm(2n+1)\frac{\pi}{2}$, where n ∈ **Z** (set of integers)

Solution-5:

$$\cos 0° = 1, \qquad \sin 0° = 0;$$

$$\cos \frac{\pi}{2} = 0, \qquad \sin \frac{\pi}{2} = 1;$$

$$\cos \pi = -1, \qquad \sin \pi = 0;$$

Solution-6:

$$cos \ \frac{3\pi}{2} = 0, \qquad sin \ \frac{3\pi}{2} = -1;$$

$$cos \ 2\pi = 1, \qquad sin \ 2\pi = 0;$$

Solution-7: Sign of function sin, cos, tan in four quadrants:

- All are positive in Q I.
- sin is positive in Q II
- tan is positive in Q III
- cos is positive in Q IV

Trigger: ASTC - All (Q-I), Sin(Q-II), Tan(Q-III), Cos(Q-IV)

Solution-8:

- sin(30°) : Positive
- cos(30°) : Positive
- tan(30°) : Positive
- sin(150°) : Positive
- cos(150°) : Negative
- tan(150°): Negative
- sin(210°) : Negative
- cos(210°) : Negative
- tan(210°) : Positive
- sin(330°) : Negative
- cos(330°) : Positive
- tan(330°) : Negative

3.3   Values of Trigonometric Functions for any angles

Now we will discuss how to find the value of trigonometric function of any angle.

To find the value of any angle in any quadrant, there are two steps:

i) We first transform the given angle into corresponding acute angle, also called reference angle.

ii) Thereafter, we apply the positive or negative sign to the value of trigonometric functions based on the quadrant, as discussed in previous section.

Reference angle/corresponding acute angle: The corresponding acute angle is the shortest angular distance to the positive or negative X-axis. The corresponding acute angle is often called the reference angle.

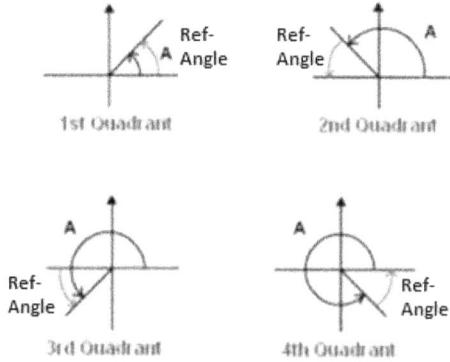

We will understand the reference angle or corresponding acute angle with following examples.

Example-1: If $\theta = 120°$ (second quadrant), then find the corresponding acute angle.

Solution: An angle of 120° has shortest possible angular distance of 60° with negative X-axis. Therefore, 60° is the corresponding acute angle.

Example-2: If $\theta = 340°$ (fourth quadrant), then find the corresponding acute angle.

Solution: An angle of 340° has shortest possible angular distance of 20° with positive X-axis. Therefore, 20° is the corresponding acute angle.

Example-3: If $\theta = 190°$ (third quadrant), then find the corresponding acute angle.

Solution: An angle of 190° has shortest possible angular distance of 10° with negative X-axis. Therefore, 10° is the corresponding acute angle.

Apply the sign based on the quadrant: We will now apply the sign based on the quadrants to the value of corresponding acute angle. Let us see following examples.

Example-1: Evaluate tan 120°

Solution: The corresponding acute angle of 120° is 60°. As 120° is in the second quadrant, the sign of tan is negative.

Therefore, $\tan 120° = -\tan 60° = -\sqrt{3}$

Example-2: Evaluate cos 210°.

Solution: The corresponding acute angle of 210° is 30°. As 210° is in the third quadrant, the sign of cos is negative.

Therefore, $\cos 210° = -\cos 30° = -\dfrac{\sqrt{3}}{2}$

Example-3: Evaluate sin 120°.

Solution: The corresponding acute angle of 120° is 60°. As 120° is in the second quadrant, the sign of sin is positive.

Therefore, $\sin 120° = +\sin 60° = \dfrac{\sqrt{3}}{2}$

Theorem 3.3: A trigonometric function of any angle will equal plus or minus of that same function of the corresponding acute angle. The sign will depend on the quadrant.

Illustrating the theorem of reference/corresponding acute angle

Assume a ray of length r revolves around circle passing through different quadrants, as depicted in following figure.

In the first quadrant with ray terminating at any (a, b), all the angles are also its reference angles. So, angle θ equals corresponding acute angle β. So, values of all the trigonometric functions are positive.

If θ is in the second quadrant, then the ray terminates at any point $(-a, b)$ and the corresponding acute angle is β, as shown in the following figure.

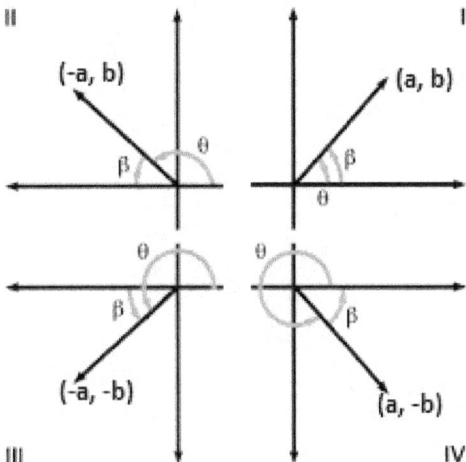

In the second quadrant,

$$\sin \theta = \frac{b}{r} = \sin \beta \qquad \textit{(value of sin is positive)}$$

while in first quadrant also, we have $\sin \theta = \frac{b}{r}$, therefore the Sine of θ is equal to the Sine of the corresponding acute angle. Similarly, other trigonometric functions are:

$$\cos \theta = \frac{-a}{r} = - \cos \beta \qquad \textit{(value of cos is negative)}$$

$$\tan \theta = \frac{b}{-a} = - \tan \beta \qquad \textit{(value of tan is negative)}$$

So, in the second quadrant, a function of θ is plus or minus that same function of reference angle β.

Next, if θ is in the third quadrant, so that r terminates at any $(-a, -b)$, then

$$\sin \theta = \frac{-b}{r} = - \sin \beta \qquad \textit{(value of sin is negative)}$$

$$\cos \theta = \frac{-a}{r} = -\cos \beta \qquad \textit{(value of cos is negative)}$$

$$\tan \theta = \frac{-b}{-a} = \tan \beta \qquad \textit{(value of tan is positive)}$$

In the third quadrant also, each function of θ is plus or minus that same function of β.

Finally, if θ is in the fourth quadrant, so that r terminates at any $(a, -b)$, then

$$\sin \theta = \frac{-b}{r} = -\sin \beta \qquad \textit{(value of sin is negative)}$$

$$\cos \theta = \frac{a}{r} = \cos \beta \qquad \textit{(value of cos is positive)}$$

$$\tan \theta = \frac{-b}{a} = -\tan \beta \qquad \textit{(value of tan is negative)}$$

In the fourth quadrant also, each function of θ is plus or minus that same function of β.

Hence, we have verified that magnitude of value remains same as corresponding acute angle for all trigonometric functions. Therefore, in every case, a function of θ is plus or minus that same function of reference angle β.

Sign & Value for trigonometric functions of any angles:

A trigonometric function of any angle will equal plus or minus of that same function of the corresponding acute angle. The sign will depend on the quadrant.

3.4 Trigonometric Functions for Negative Angles

Let's consider a revolving line making different angles θ. No matter in which quadrant the revolving line terminates, both positive and negative angles (θ and $-\theta$) have the same corresponding acute angle (or reference angle). Moreover, both angles on the X-Y plane fall in the same left-or-right half of the Y-axis.

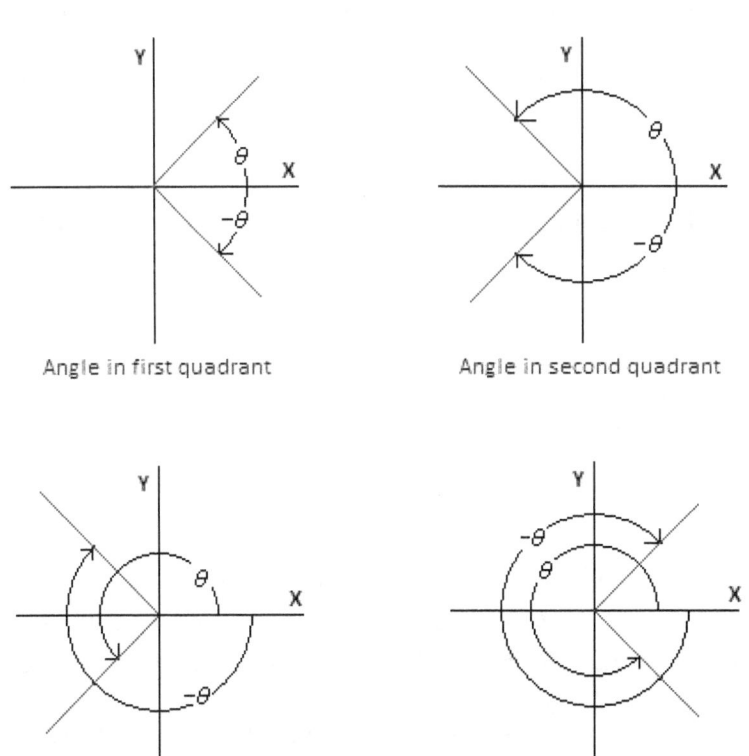

Angle in first quadrant   Angle in second quadrant

Angle in third quadrant   Angle in fourth quadrant

In right half (first and fourth quadrants) the Cosine is positive but in left half (second and third quadrants) it is negative. This implies that the sign of the Cosine depends only on which left or right half its angle terminates. But θ and $-\theta$ also falls together on the same right or left half of Y-axis. Therefore, cos $(-\theta)$ and cos θ will have the same sign:

$$\cos(-\theta) = \cos\theta$$

On the other hand, angles θ and $-\theta$ on the X-Y plane fall in opposite top-or-bottom halves of the X-axis. The sign of the Sine is positive on top half and negative on bottom half. Therefore, sin $(-\theta)$ and sin θ will have opposite signs:

$$\sin(-\theta) = -\sin\theta$$

As the Tangent is quotient of Sine and Cosine, tan $(-\theta)$ and tan θ will have opposite sign.

$$\tan(-\theta) = -\tan\theta$$

Note that values of tan θ are negative for $\pi/2 < \theta < \pi$; and $3\pi/2 < \theta < 2\pi$.

In the terms of odd/even, Cosine is an even function, because **cos(−θ)=cos θ** *. But Sine is an odd function, because* **sin(−θ)=−sin θ**. *Similarly, Tangent is an odd function because* **tan(−θ)=−tan θ**.

In the same way, cosine and cot are odd functions, but sec is an even function.

Example 1: Find cos $(-30°)$

Solution: cos $(-30°) = \cos 30° = \dfrac{\sqrt{3}}{2}$

Example 2: Find $\cot(-30°)$

Solution: $\cot(-30°) = -\cot 30° = -\sqrt{3}$

Example 3: Find $\operatorname{cosec}(-45°)$

Solution: $\operatorname{cosec}(-45°) = -\operatorname{cosec} 45° = -\sqrt{2}$

Negative identities:

$\sin(-\theta) = -\sin\theta;$ $\operatorname{cosec}(-\theta) = -\operatorname{cosec}\theta;$

$\cos(-\theta) = \cos\theta;$ $\sec(-\theta) = \sec\theta;$

$\tan(-\theta) = -\tan\theta;$ $\cot(-\theta) = -\cot\theta;$

Worksheet-8

Exercise-1: Find the corresponding acute angle (reference angle) for following angles:

a) 110°

b) 225°

c) −30°

d) 380°

Exercise-2: Evaluate tan 120°.

Exercise-3: Evaluate cos 195°.

Exercise-4: Evaluate sec (−45°).

Exercise-5:   Evaluate each of the following without referring tables.

    a) sin 150°

    b) cos 135°

c) cosec (−30)°

d) tan 240°

Eercise-6: Use negative identity to evaluate sin (−45°)

Exercise-7: Use negative identity to evaluate cot (−30°)

Exercise-8: Use negative identity to evaluate sec (−30°)

Exercise-9: Apply negative identity to evaluate the following:

a) cos (−30°)

b) cos (−60°)

c) cos (−45°)

d) sin (−30°)

e) sin (−60°)

f) sin (−45°)

Exercise-10: If $\theta = \dfrac{5\pi}{6}$, find exact values for sec θ, cosec θ, tan θ, cot θ.

Exercise-11: If $\theta = \dfrac{2\pi}{3}$, find exact values for sec θ, cosec θ, tan θ, cot θ.

Exercise-12: Evaluate: a. sec (135°) b. cosec (210°)   c. tan (60°)   d. cot (225°)

Exercise-13: If $\sin \theta = \dfrac{3}{4}$, and θ is in quadrant II, find cos θ, sec θ, cosec θ, tan θ, cot θ.

Exercise-14: If $\cos \theta = -\dfrac{1}{3}$, and θ is in quadrant III, find $\sin \theta$, $\sec \theta$, $cosec\ \theta$, $tan\ \theta$, $cot\ \theta$.

Exercise-15: $\tan \theta = \dfrac{12}{5}$, and $0 \le \theta < \dfrac{\pi}{2}$, find $\sin \theta$, $\cos \theta$, $\sec \theta$, $cosec\ \theta$, $cot\ \theta$.

Solutions-Worksheet-8

Solution-1:

1-a: $70°$;   1-b: $45°$;   1-c: $30°$;   1-d: $20°$

Solution-2: The corresponding acute angle of $120°$ is $60°$. Therefore, according to the theorem,

$$\tan 120° => \pm \tan 60° = \pm \sqrt{3}$$

As $120°$ is in the second quadrant, $\tan 120°$ is negative. Therefore,

$$\tan 120° = - \sqrt{3}$$

Solution-3: The corresponding acute angle of $195°$ (third quadrant) is $15°$. Therefore,

$\cos 195° => \pm\cos 15° = \pm\, 0.966$ (Refer the table in appendix-1)

As $195°$ is in third quadrant, $\cos 195°$ is negative. Therefore,

$\cos 195° = -\, 0.966$

Solution-4: The corresponding acute angle of $-45°$ (fourth quadrant) is $45°$. Therefore,

$\sec -45° => \pm\sec 45° = \pm \sqrt{2}$

In the fourth quadrant, $\sec -45°$ is positive. Therefore,

$\sec -45° = \sqrt{2}$

Solution-5:

Solution-5-a: The corresponding acute angle is $30°$. As $150°$ is in the second quadrant, sin 150° is positive. Therefore,

$$\sin 150° = \sin 30° = \frac{1}{2}$$

Solution-5-b: The corresponding acute angle is 45°. As 135° is in the second quadrant, cos 135° is negative. Therefore,

$$\cos 135° = -\cos 45° = -\frac{1}{\sqrt{2}}$$

Solution-5-c: The corresponding acute angle is 30°. As −30° is in the fourth quadrant, cosec −30° is negative. Therefore,

$$\text{cosec } (-30)° = -\text{cosec } 30° = -2.$$

Solution-5-d: The corresponding acute angle is 60°. As 240° is in the third quadrant, tan 240° is positive. Therefore,

$$\tan 240° = \tan 60° = \sqrt{3}$$

Solution-6: $\sin(-45°) = -\sin 45° = -\dfrac{1}{\sqrt{2}}$

Solution-7: $\cot(-30°) = -\cot 30° = -\sqrt{3}$

Solution-8: $\sec(-30°) = \sec 30° = \dfrac{2}{\sqrt{3}}$

Solution-9:

a: $\cos(-30°) = \cos 30° = \dfrac{\sqrt{3}}{2}$

b: $\cos(-60°) = \cos 60° = \dfrac{1}{2}$

c: $\cos(-45°) = \cos 45° = \dfrac{1}{\sqrt{2}}$

d: $\sin(-30°) = -\sin 30° = -\dfrac{1}{2}$

e: $\sin(-60°) = -\sin 60° = -\dfrac{\sqrt{3}}{2}$

f: $\sin(-45°) = -\sin 45° = -\dfrac{1}{\sqrt{2}}$

Solution-10: The corresponding acute angle for $\dfrac{5\pi}{6}$ is $\dfrac{\pi}{6}$. As $\dfrac{5\pi}{6}$ (or 150°) is in the second quadrant, sin 150° is positive. Therefore,

$$\sin 150° = \sin 30° = \dfrac{1}{2}$$

In the second quadrant, cos is negative. Therefore,

$$\cos 150° = -\cos 30° = -\dfrac{\sqrt{3}}{2}$$

So, we can calculate the following with the value of Sine and Cosine.

$$\sec\left(\dfrac{5\pi}{6}\right) = \dfrac{1}{\cos\left(\dfrac{5\pi}{6}\right)} = \dfrac{1}{-\dfrac{\sqrt{3}}{2}} = -\dfrac{2}{\sqrt{3}} = -\dfrac{2\sqrt{3}}{3}$$

$$\csc\left(\dfrac{5\pi}{6}\right) = \dfrac{1}{\sin\left(\dfrac{5\pi}{6}\right)} = \dfrac{1}{\dfrac{1}{2}} = 2$$

$$\tan\left(\dfrac{5\pi}{6}\right) = \dfrac{\sin\left(\dfrac{5\pi}{6}\right)}{\cos\left(\dfrac{5\pi}{6}\right)} = \dfrac{\dfrac{1}{2}}{\dfrac{-(\sqrt{3})}{2}} = -\dfrac{1}{\sqrt{3}}$$

$$\cot\left(\dfrac{5\pi}{6}\right) = \dfrac{1}{\tan\left(\dfrac{5\pi}{6}\right)} = \dfrac{1}{-\dfrac{\sqrt{3}}{3}} = -\sqrt{3}$$

Solution-11: The corresponding acute angle for $\dfrac{2\pi}{3}$ is $\dfrac{\pi}{3}$. As $\dfrac{2\pi}{3}$ (or 120°) is in the second quadrant, sin is positive. Therefore,

$$\sin 120° = \sin 60° = \dfrac{\sqrt{3}}{2}$$

In the second quadrant, cos is negative. Therefore,

$$\cos 120° = -\cos 60° = -\frac{1}{2}$$

Applying the values of Sine and Cosine to find other values, we have

$$\sec\left(\frac{2\pi}{3}\right) = \frac{1}{\cos\left(\frac{2\pi}{3}\right)} = \frac{1}{\frac{-1}{2}} = -2$$

$$\operatorname{cosec}\left(\frac{2\pi}{3}\right) = \frac{1}{\sin\left(\frac{2\pi}{3}\right)} = \frac{1}{\frac{\sqrt{3}}{2}} = \frac{2\sqrt{3}}{3};$$

$$\tan\left(\frac{2\pi}{3}\right) = \frac{\sin\left(\frac{2\pi}{3}\right)}{\cos\left(\frac{2\pi}{3}\right)} = \frac{\frac{\sqrt{3}}{2}}{\frac{-1}{2}} = -\sqrt{3}$$

$$\cot\left(\frac{2\pi}{3}\right) = \frac{1}{\tan\left(\frac{2\pi}{3}\right)} = \frac{1}{-\sqrt{3}} = -\frac{\sqrt{3}}{3}$$

Solution-12:

a-The corresponding acute angle for 135° (second quadrant) is 45°, and Cosine is negative in the second quadrant. Therefore,

$$\sec 135° = \frac{1}{\cos(135°)} = \frac{1}{-\cos(45°)} = \frac{1}{\frac{-\sqrt{2}}{2}} = -\sqrt{2}$$

b-The corresponding acute angle for 210° (third quadrant) is 30°, and Sine is negative in the third quadrant. Therefore,

$$\operatorname{cosec} 210° = \frac{1}{\sin(210°)} = \frac{1}{-\sin(30°)} = \frac{1}{\frac{-1}{2}} = -2$$

c-Transform tan into sin and cos; $\tan 60° = \dfrac{\sin(60°)}{\cos(60°)} = \dfrac{\frac{\sqrt{3}}{2}}{\frac{1}{2}} = \sqrt{3}$

d-The corresponding acute angle for $225°$ (third quadrant) is $45°$, and tan is positive in third quadrant. Therefore,

$$\cot 225° = \frac{1}{\tan(225°)} = \frac{1}{\tan(45°)} = \frac{1}{1} = 1$$

Solution-13: Because θ is in second quadrant, so the signs of functions are: $\cos(\theta) < 0, \sec(\theta) < 0;\ \sin(\theta) > 0, \operatorname{cosec}(\theta) > 0;\ \tan(\theta) < 0, \cot(\theta) < 0$. Therefore,

$$\cos(\theta) = -\sqrt{1 - \sin^2(\theta)} = -\sqrt{1 - \left(\frac{3}{4}\right)^2} = -\frac{\sqrt{7}}{4}$$

$$\sec \theta = \frac{1}{\cos(\theta)} = \frac{1}{\frac{-\sqrt{7}}{4}} = -\frac{4}{\sqrt{7}}$$

$$\operatorname{cosec} \theta = \operatorname{cosec}(\theta) = \frac{1}{\sin(\theta)} = \frac{1}{\frac{3}{4}} = \frac{4}{3}$$

$$\tan \theta = \frac{\sin(\theta)}{\cos(\theta)} = \frac{\frac{3}{4}}{\frac{-\sqrt{7}}{4}} = \frac{-3}{\sqrt{7}};$$

$$\cot \theta = \frac{1}{\tan \theta} = \frac{1}{\frac{-3\sqrt{7}}{7}} = \frac{7}{-3\sqrt{7}} = -\frac{\sqrt{7}}{3}$$

Solution-14: Because θ is in third quadrant, we have $\cos(\theta) < 0, \sec(\theta) < 0;\ \sin(\theta) < 0, \operatorname{cosec}(\theta) < 0;\ \tan(\theta) > 0, \cot(\theta) > 0$. Therefore,

$$\sin(\theta) = -\sqrt{1 - \cos^2(\theta)} = \frac{-2\sqrt{2}}{3}; \ \csc(\theta) = \frac{1}{\sin(\theta)} =$$

$$\frac{3}{-2\sqrt{2}} = -\frac{3\sqrt{2}}{4}; \sec(\theta) = \frac{1}{\cos(\theta)} = -3; \tan(\theta) = \frac{\sin(\theta)}{\cos(\theta)} =$$

$$2\sqrt{2}; \ \cot(\theta) = \frac{1}{\tan(\theta)} = \frac{1}{2\sqrt{2}} = \frac{\sqrt{2}}{4}$$

Solution-15:

$0 \le \theta \le \frac{\pi}{2}$ means θ is in the first quadrant, so all trigonometric ratios are positive.

Since $\quad \tan(\theta) = \dfrac{perpendicular}{base} = \dfrac{12}{5}$, we can apply the Pythagorean formula to find hypotenuse using the perpendicular be 12 and base be 5. Hypotenuse r is calculated as:

$$r = \sqrt{12^2 + 5^2} = 13$$

Therefore,

$$\sin(\theta) = \frac{12}{13}; \csc(\theta) = \frac{1}{\sin(\theta)} = \frac{13}{12}; \cos(\theta) = \frac{5}{13};$$
$$\sec(\theta) = \frac{1}{\cos(\theta)} = \frac{13}{5}; \cot(\theta) = \frac{1}{\tan(\theta)} = \frac{5}{12}$$

3.5 Domain & Range of Trigonometric Functions

Let's consider a function, $y = \sin x$

In this function, range is all possible values of y, and domain is all possible values of x.

*We use the set definition for the range and domain of trigonometric functions. In this book, we use the commonly used symbols, like "**R**" as a set of all real number and "**Z**" as a set of all integers.*

a- Domain and range of sin x and cos x

For each angle x, $-1 < \sin x < 1$, and $-1 < \cos x < 1$

Therefore,

Domain of function, $y = \sin x$ is the set of all real numbers and range is the interval $[-1, 1]$ because $-1 \le y \le 1$

Domain of function, $y = cos\ x$ is the set of all real numbers and range is the interval $[-1, 1]$ because $-1 \le y \le 1$

b- Domain and range of cosec x and sec x

Since cosec $x = 1/\sin x$, so

Domain of function $y = \text{cosec } x$ is the set $\{x : x \in \mathbf{R} \text{ and } x \neq n\pi,\ n \in \mathbf{Z}\}$ and range is the set $\{y : y \in \mathbf{R}, y \ge 1 \text{ or } y \le -1\}$

Note that for the function y = cosec x; x ≠ nπ, otherwise y will be undefined.

Similarly, since sec x = 1/cos x, so

domain of function y = sec x is the set $\{x : x \in \mathbf{R} \text{ and } x \neq (2n+1)\frac{\pi}{2},$ $n \in \mathbf{Z}\}$ and range is the set $\{y : y \in \mathbf{R}, y \leq -1 \text{ or } y \geq 1\}$

Note that for the function y = sec x,   x ≠ (2n+1) $\frac{\pi}{2}$, otherwise y will be undefined.

c- Domain and range of tan x and cot x

Domain of y = tan x is the set $\{x : x \in \mathbf{R} \text{ and } x \neq (2n+1)\frac{\pi}{2}, n \in \mathbf{Z}\}$ and range is the set of all real numbers.

Note that for the function y = tan x,   x ≠ (2n+1) $\frac{\pi}{2}$, otherwise y will be undefined.

Domain of y = cot x is the set $\{x : x \in \mathbf{R} \text{ and } x \neq n\pi, n \in \mathbf{Z}\}$ and range is the set of all real numbers.

Note that for function y = cot x,   x ≠ nπ, otherwise y will be undefined.

Summary note for range of all trigonometric functions in different quadrants:

	Quadrant I	Quadrant II	Quadrant III	Quadrant IV
sin	increases from 0 to 1	declines from 1 to 0	declines from 0 to −1	increases from −1 to 0
cos	declines from 1 to 0	declines from 0 to −1	increases from −1 to 0	increases from to 0 to 1
tan	increases from 0 to ∞	increases from −∞ to 0	increases from 0 to ∞	increases from −∞ to 0
cot	declines from ∞ to 0	declines from 0 to −∞	declines from ∞ to 0	declines from 0 to −∞
sec	increases from 1 to ∞	increases from −∞ to −1	declines from −1 to −∞	declines from ∞ to 1
cosec	declines from ∞ to 1	increases from 1 to ∞	increases from −∞ to −1	declines from −1 to −∞

This table should be understood and memorized to be proficient in higher level trigonometry.

In the above table, *sin x* increases from 0 to 1 in quadrant I implies that *sin x* increases as *x* increases from 0 to $\pi/2$. In quadrant II, *sin x* decreases from 1 to 0, implies that *sin x* decreases as *x* increases from $\pi/2$ to π.

Similar explanations apply for other trigonometric functions in different quadrants. The symbol "∞" represents infinity which is very large value in positive direction and " − ∞" represents minus infinity which is very large value in negative direction.

3.6 Graphs of Trigonometric Functions

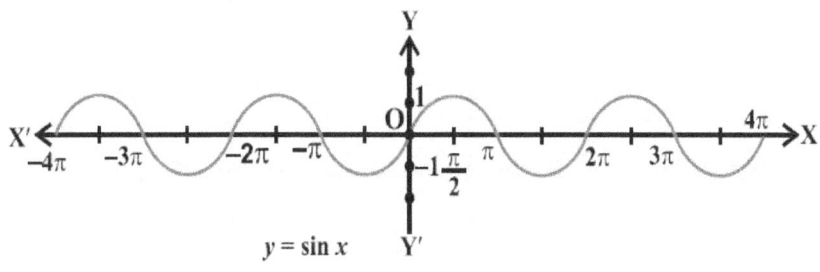

$y = \sin x$

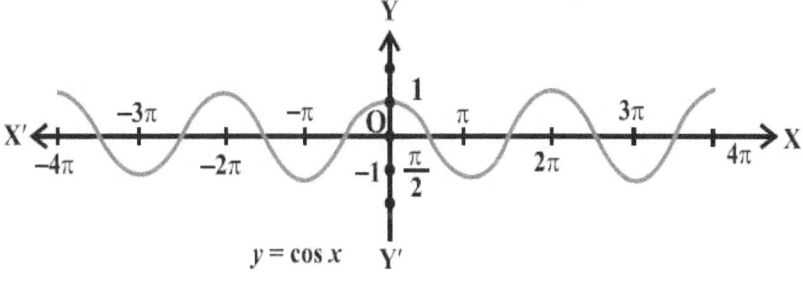

$y = \cos x$

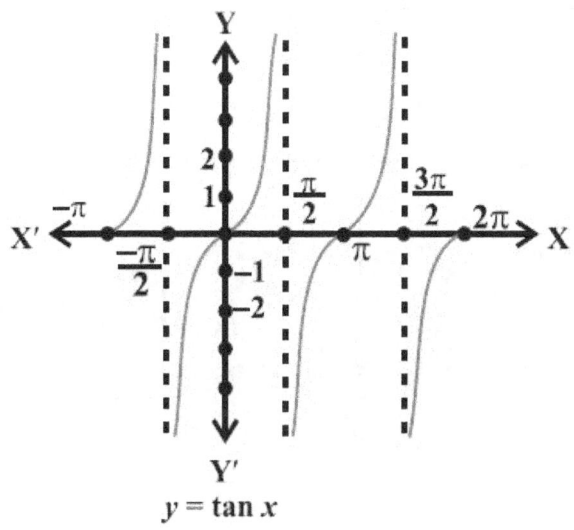

$y = \tan x$

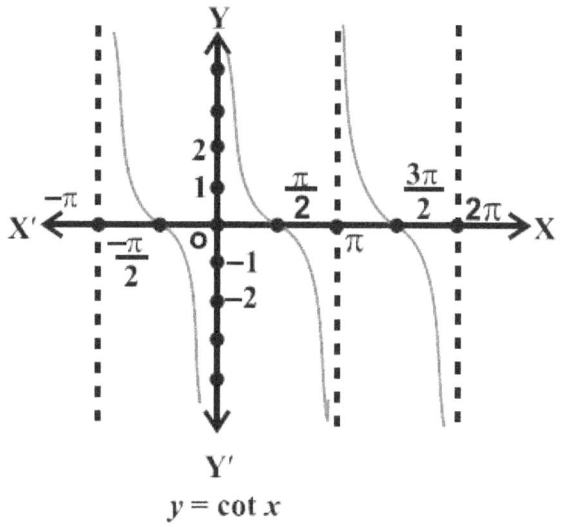

$y = \cot x$

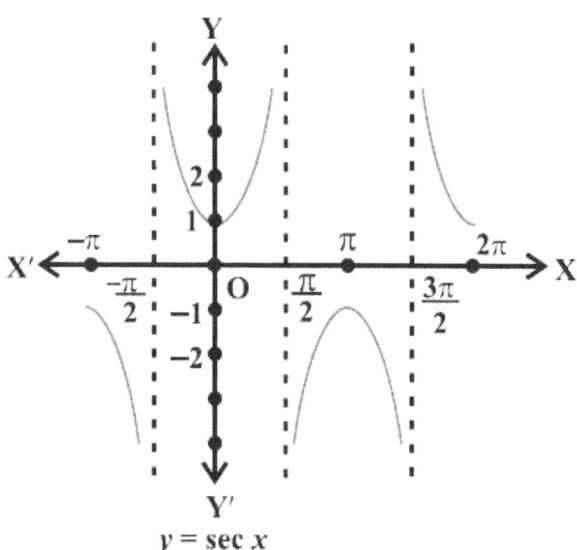

$y = \sec x$

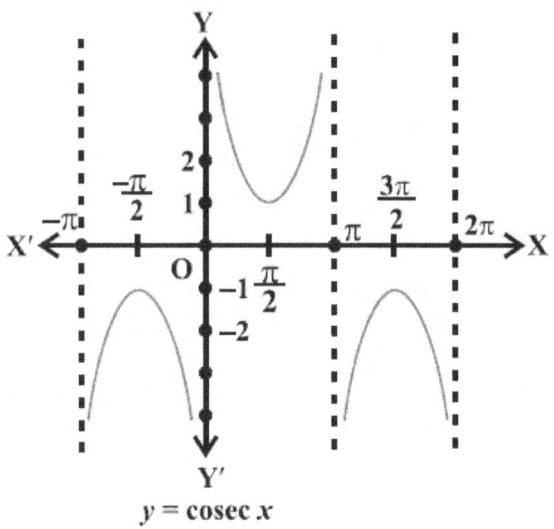

$y = \operatorname{cosec} x$

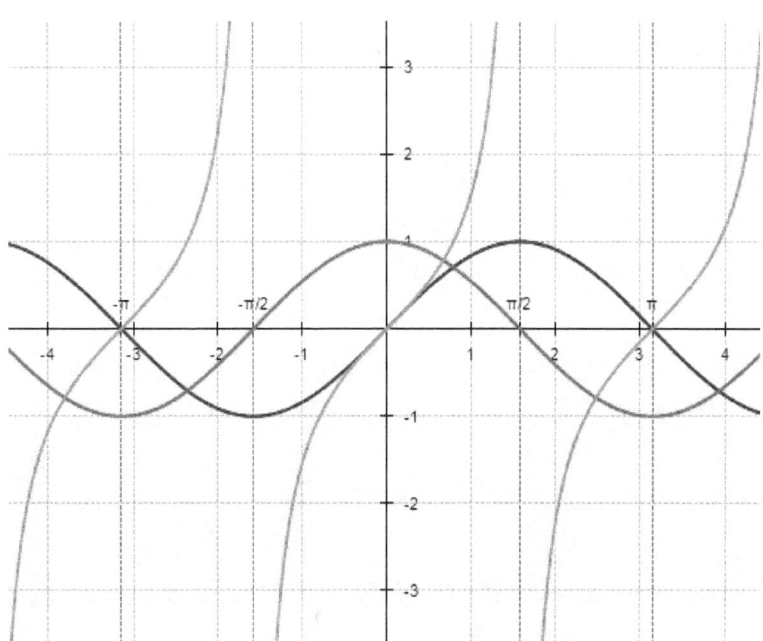

Combining the graphs of sin, cos & tan together

Worksheet-9

Exercise-1: Write the range and domain of sin x and cos x

Exercise-2: Write the range and domain of tan x and cot x

Exercise-3: Write the range and domain of sec x and cosec x

Exercise-4: Draw graph of sin x and cos x

Exercise-5: Draw graph of tan x and cot x

Exercise-6: Draw graph of cosec x and sec x

Solutions-Worksheet-9

For solutions of Exercise-1 to Exercise-6, refer the sections 3.5 and 3.6 of this chapter.

4 Chapter 4: Trigonometric Identities and Formulas

4.1   Trigonometric Identities

Trigonometric identities are equalities of trigonometric functions which are true for every single value of the occurring variables and have been standardized as identities.

4.1.1   Negative Identities

Trigonometrical ratios of an angle $-\theta$ can be transformed into positive angle θ. The details are elaborated in section 3.4 of last chapter.

Negative Identities:

$sin\ (-\theta) = -sin\ \theta$

$cos\ (-\theta) = cos\ \theta$

$tan\ (-\theta) = -tan\ \theta$

$cot\ (-\theta) = -cot\ \theta$

$sec\ (-\theta) = sec\ \theta$

$cosec\ (-\theta) = -cosec\ \theta$

4.1.2   Complementary Identities of $(\pi/2 - \theta)$

Two angles are said to be complementary when their sum is equal to a right angle i.e. 90°. So, complementary angle of any angle θ is $90° - \theta$.

Example-1: Find complementary angle of 60°

Solution: Complementary angle of 60° = 90° − 60° = 30°

Example-2: Find complementary angle of −30°

Solution: Complementary angle of −30° = 90°− (−30°) = 120°

Identities of complementary angles

Trigonometric function of angle 90°−θ can be transformed into angle θ.

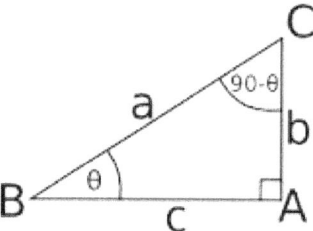

In any right triangle, as depicted in figure, we will look at the angle θ and its complementary angle 90°− θ to derive the complementary angle identities.

The value of sin with the respect to angle 90°− θ is:

$$\sin (90°-\theta) = \frac{BA}{BC}$$

But the value of cos with the respect to complementary angle θ is also:

$$\cos \theta = \frac{BA}{BC}$$

So, with respect to complementary angle sin (90°− θ) equals cos θ. Similarly, we have other complementary identities (also called cofunction identities) as summarized below.

Complementary Identities / Cofunction Identities:

$sin\,(90° - \theta) = cos\,\theta$

$cos\,(90° - \theta) = sin\,\theta$

$tan\,(90° - \theta) = cot\,\theta$

$cot\,(90° - \theta) = tan\,\theta$

$sec\,(90° - \theta) = cosec\,\theta$

$cosec\,(90° - \theta) = sec\,\theta$

Remarks: sin of any angle is cos of its complimentary angle; tan of any angle is cot of its complimentary angle; and sec of any angle is cosec of its complementary angle.

4.1.3 Trigonometric Identities of ($\pi/2 + \theta$)

Trigonometric function of angle (90°+θ) can be transformed into angle θ. The trigonometrical ratios are same as complementary identities but the sign is used for second quadrant of corresponding function.

Identities of (90°+θ)

$sin\,(90° + \theta) = cos\,\theta$

$cos\,(90° + \theta) = -sin\,\theta$

$tan\,(90° + \theta) = -cot\,\theta$

$cot\,(90° + \theta) = -tan\,\theta$

$sec\,(90° + \theta) = -cosec\,\theta$

$cosec\,(90° + \theta) = sec\,\theta$

4.1.4   Supplementary Identities of $(\pi-\theta)$

Two angles are said to supplementary if their sum is equal to two right angles, i.e. 180°. So, supplementary angle of any angle θ is $(180°-\theta)$.

Example-1: Find supplementary angle of 60°

Solution: Supplementary angle of 60° = 180°− 60° = 120°

Example-2: Find supplementary angle of −30°

Solution: Supplementary angle of −30° = 180°− (−30°) = 210°

Identities of supplementary angles

Trigonometric function of angle 180°−θ can be transformed into angle θ.

Illustration: The Sine of an obtuse angle is equal to the Sine of its supplementary angle. The Cosine of an obtuse angle is equal to the negative of the Cosine of its supplementary angle. The Tangent of an obtuse angle is equal to the negative of the Tangent of its supplementary angle.

Refer the figure depicted below.

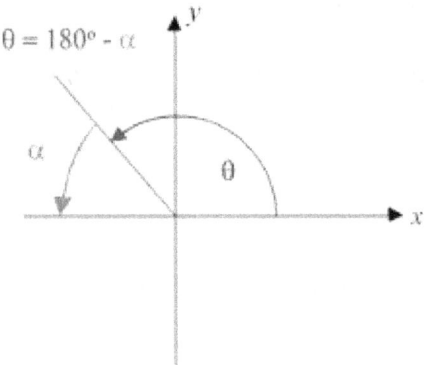

The revolving line terminates in second quadrant with an angle θ.

The supplementary angle of θ is angle $(180° - \theta)$ or angle α, which is also corresponding acute angle (reference angle) .

Thereupon, we apply the sign of second quadrant for trigonometric function.

The sign of sin in second quadrant is positive. Therefore,
$$\sin (180° - \theta) = \sin \theta$$
The sign of cos in second quadrant is negative. Therefore,
$$\cos (180° - \theta) = - \cos \theta$$
The sign of tan in second quadrant is negative. Therefore,
$$\tan (180° - \theta) = - \tan \theta$$
The trigonometric function cosec possesses the same sign as sin, so
$$\operatorname{cosec} (180° - \theta) = \operatorname{cosec} \theta$$
The trigonometric function sec possesses the same sign as cos, so
$$\sec (180° - \theta) = - \sec \theta$$
The trigonometric function cot possesses the same sign as tan, so
$$\cot (180° - \theta) = - \cot \theta$$

Alternatively, we can verify identities of $(180°-\theta)$ using Identities of $(90°+\theta)$

$\sin (180° - \theta) = \sin (90° + (90° - \theta)) = \cos (90° - \theta) = \sin \theta$

$\cos (180° - \theta) = \cos (90° + (90° - \theta)) = -\sin (90° - \theta) = -\cos \theta$

$\tan (180° - \theta) = \tan (90° + (90° - \theta)) = -\cot (90° - \theta) = -\tan \theta$

Supplementary Identities of $(180°-\theta)$:

$\sin (180° - \theta) = \sin \theta$

$\cos (180° - \theta) = -\cos \theta$

$\tan (180° - \theta) = -\tan \theta$

$\cot (180° - \theta) = -\cot \theta$

$\sec (180° - \theta) = -\sec \theta$

$\operatorname{cosec} (180° - \theta) = \operatorname{cosec} \theta$

4.1.5   Trigonometric Identities of $(\pi+\theta)$

Trigonometric function of angle $180°+\theta$ can be transformed into angle θ.

For given angle $(180°+ \theta)$, angle θ is corresponding acute angle or reference angle. The trigonometric function remains same. Thereupon, we apply the sign for trigonometric function in third quadrant.

sin is negative in third quadrant, so $\sin (180°+ \theta) = -\sin \theta$

cos is negative in third quadrant, so $\cos (180°+ \theta) = -\cos \theta$

tan is positive in third quadrant, so $\tan (180°+ \theta) = \tan \theta$

Alternatively, we can verify identities of (180°+θ) using Identities of (90°+θ);

$\sin (180°+ \theta) = \sin (90°+ (90° + \theta)) = \cos (90°+ \theta) = -\sin \theta$

$\cos (180°+ \theta) = \cos (90°+ (90° + \theta)) = -\sin (90°+ \theta) = -\cos \theta$

$\tan (180°+ \theta) = \tan (90°+ (90° + \theta)) = -\cot (90°+ \theta) = \tan \theta$

Remarks: The trigonometrical identities of (180°+ θ) maintain the same trigonometric function but sign is used of third quadrant.

Illustration: $\cos (\pi + \theta) = -\cos \theta$

As depicted in the figure below, no matter in which quadrant θ falls, θ and $(\pi + \theta)$ will have the same corresponding acute angle.

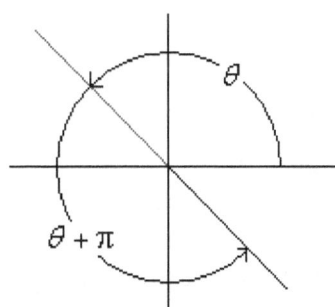

As shown, $(\pi + \theta)$ will fall in the opposite left-or-right half of the plane. We know that cos is positive in right half of circle (1st and 4th quadrants) but negative in left half (2nd and 3rd quadrants).

Therefore, $\cos \theta$ and $\cos (\pi + \theta)$ will have opposite sign:

$\cos (\pi + \theta) = -\cos \theta.$

Similarly, the identities of (π + θ) for sin and tan can be illustrated. It is to be noticed that in upper half of circle (1st and 2nd quadrants), sin is positive. In lower half of circle (3rd and 4th quadrants), sin is negative.

Identities of 180°+θ:

$sin\ (180°+ \theta) = -sin\ \theta$
$cos\ (180°+ \theta) = -cos\ \theta$
$tan\ (180°+ \theta) = tan\ \theta$
$cot\ (180°+ \theta) = cot\ \theta$
$sec\ (180°+ \theta) = -sec\ \theta$
$cosec\ (180°+ \theta) = -cosec\ \theta$

4.1.6   Trigonometric Identities of (2π−θ)

Trigonometric function of angle $(2\pi - \theta)$ can be transformed into angle θ.

For given angle (360°− θ), angle θ is corresponding acute angle or reference angle. The trigonometric function remains same. Thereupon, we apply the sign for trigonometric function in fourth quadrant.

sin is negative in fourth quadrant, so sin (360°− θ) = −sin θ

cos is positive in fourth quadrant, so cos (360°− θ) = cos θ

tan is negative in fourth quadrant, so tan (360°− θ) = −tan θ

Alternatively, we can verify identities of (360°–θ) using Identities of (180°+θ) and (180°–θ);

$\sin(360° - \theta) = \sin(180° + (180° - \theta)) = -\sin(180° - \theta) = -\sin\theta$

$\cos(360° - \theta) = \cos(180° + (180° - \theta)) = -\cos(180° - \theta) = \cos\theta$

$\tan(360° - \theta) = \tan(180° + (180° - \theta)) = \tan(180° - \theta) = -\tan\theta$

Similarly, for other reciprocal functions, we have:

$\cot(360° - \theta) = -\cot\theta$
$\sec(360° - \theta) = \sec\theta$
$\csc(360° - \theta) = -\csc\theta$

Identities of 360°–θ:

$\sin(360° - \theta) = -\sin\theta$
$\cos(360° - \theta) = \cos\theta$
$\tan(360° - \theta) = -\tan\theta$
$\cot(360° - \theta) = -\cot\theta$
$\sec(360° - \theta) = \sec\theta$
$\csc(360° - \theta) = -\csc\theta$

4.1.7   Trigonometric Identities of $(2\pi+\theta)$

One complete revolution makes an angle 2π or $360°$. Any revolving line describing an angle θ will be in exactly same position when it makes one or many complete revolutions in clockwise or counter-clockwise direction. So angles θ and $(2\pi + \theta)$ of any trigonometric function will have same value and sign. These angles are also called coterminal angles.

So, trigonometrical ratios of angle (360°+θ) maintain same value and sign as angle θ.

Identities of 360°+θ:

sin (360°+ θ) = sin θ

cos (360°+ θ) = cos θ

tan (360°+ θ) = tan θ

cot (360°+ θ) = cot θ

sec (360°+ θ) = sec θ

cosec (360°+ θ) = cosec θ

Remarks: With elaboration of angle transformation mentioned above, a trigonometrical function of any angle can be reduced to trigonometrical function of an angle lying between 0° and 45°

4.2 Trigonometric Identities for Sum-Difference of Angles

Sum and Difference Identities:

Sum and difference of angles are very useful in trigonometry and these formulas are treated as identities.

$$\sin(\alpha + \beta) = \sin\alpha\cos\beta + \cos\alpha\sin\beta$$

$$\sin(\alpha - \beta) = \sin\alpha\cos\beta - \cos\alpha\sin\beta$$

$$\cos(\alpha + \beta) = \cos\alpha\cos\beta - \sin\alpha\sin\beta$$

$$\cos(\alpha - \beta) = \cos\alpha\cos\beta + \sin\alpha\sin\beta$$

Remark: These sum and difference identities of sin and cos are true for α and β be any angle.

Derived sum/difference identities for tan:

$$\tan(\alpha + \beta) = \frac{\tan\alpha + \tan\beta}{1 - \tan\alpha\tan\beta} , \ \textit{where none of the angles } \alpha, \beta$$

$$\textit{and } (\alpha + \beta) \textit{ are an odd multiple of } \frac{\pi}{2}$$

$$\tan(\alpha - \beta) = \frac{\tan\alpha - \tan\beta}{1 + \tan\alpha\tan\beta} , \ \textit{where none of the angles } \alpha, \beta$$

$$\textit{and } (\alpha - \beta) \textit{ are an odd multiple of } \frac{\pi}{2}$$

Derived sum/difference identities for cot:

$$\cot(\alpha + \beta) = \frac{\cot\alpha\cot\beta - 1}{\cot\alpha + \cot\beta}$$, *where none of the angles* α, β

and $(\alpha + \beta)$ *are a multiple of* π

$$\cot(\alpha - \beta) = \frac{\cot\alpha\cot\beta + 1}{\cot\beta - \cot\alpha}$$, *where none of the angles* α, β

and $(\alpha - \beta)$ *are a multiple of* π

Proof - Identities of Sum/Difference of Angles:

Identity of sum of angles for sin:

Let us consider a straight line OX rotating about centre O in anti-clockwise direction. At point Y, the line OX sweeps over the angle α; and from point Y to Z it sweeps the angle β as depicted in figure below.

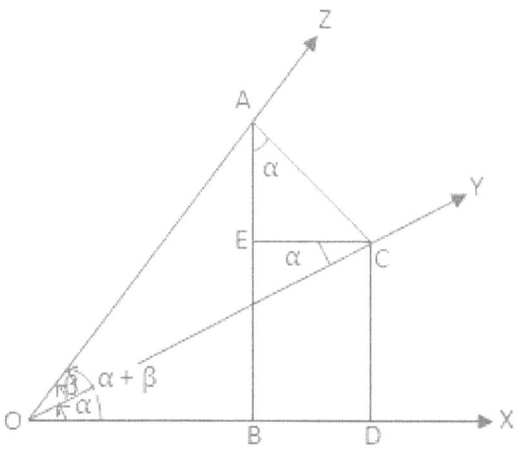

With the help of this construction, we will verify following.

$$\sin (\alpha + \beta) = \sin \alpha \cos \beta + \cos \alpha \sin \beta$$

To begin with, we draw a line AB perpendicular to OX. Further we draw lines AC perpendicular to OY, CD perpendicular to OX, and CE perpendicular to AB.

As we know the theorem that when a straight line (transversal) crosses two parallel straight lines, then the alternate angles are equal. In the figure, OC is transversal crossing two parallel lines EC and OD,

Therefore, the alternate angles $\angle \text{YOX} = \angle \text{ECO} = \alpha$

Moreover, $\angle$ ACE is complementary angle of $\angle$ ECO, and $\angle$ AEC is a right angle, so in a right triangle AEC, $\angle \text{EAC} = \alpha$

Furthermore EBDC is a rectangle, so BE = DC

$$\sin (\alpha + \beta) = \frac{AB}{AO} = \frac{BE + EA}{AO} = \frac{DC + EA}{AO} = \frac{DC}{AO} + \frac{EA}{AO}$$

Dividing and multiplying both the terms of numerator and denominator by OC and CA respectively, we get

$$= \frac{DC}{OC} \frac{OC}{AO} + \frac{EA}{CA} \frac{CA}{AO}$$

$$= \sin \alpha \cos \beta + \cos \alpha \sin \beta \qquad \text{QED.}$$

Example-1: Find the value of sin 75° using sum of angles identity of sin.

Solution: $\sin 75° = \sin (45° + 30°)$

$$= \sin 45° \cos 30° + \cos 45° \sin 30°$$

$$= \frac{1}{\sqrt{2}} \frac{\sqrt{3}}{2} + \frac{1}{\sqrt{2}} \frac{1}{2}$$

$$= \frac{\sqrt{3}}{2\sqrt{2}} + \frac{1}{2\sqrt{2}} = \frac{\sqrt{3}+1}{2\sqrt{2}}$$

Identity of difference of angles for sin:

Now we will verify the difference identities for sin. To do this, we use its sum identity besides using the appropriate sign. So difference identity is transformed into sum identity as follows:

$$\sin(\alpha - \beta) = \sin(\alpha + (-\beta))$$

Using sum identity of sin,

$$\sin(\alpha + (-\beta)) = \sin\alpha\cos(-\beta) + \cos\alpha\sin(-\beta)$$

Since $\cos(-\beta) = \cos\beta$, and $\sin(-\beta) = -\sin\beta$,

Hence,

$$\sin(\alpha - \beta) = \sin\alpha\cos\beta - \cos\alpha\sin\beta \qquad \text{QED.}$$

Example-1: Find the value of sin 15° using difference of angle identity of sin.

Solution: $\sin 15° = \sin(45° - 30°)$

$$= \sin 45° \cos 30° - \cos 45° \sin 30°$$

$$= \frac{1}{\sqrt{2}} \frac{\sqrt{3}}{2} - \frac{1}{\sqrt{2}} \frac{1}{2}$$

$$= \frac{\sqrt{3}}{2\sqrt{2}} - \frac{1}{2\sqrt{2}} = \frac{\sqrt{3}-1}{2\sqrt{2}}$$

Sum of angles identity for cos:

We will use the same construction of figure which we used to prove the identity of sum of angles for sin. Identity of sum of angles for cos is: $\cos(\alpha + \beta) = \cos\alpha\cos\beta - \sin\alpha\sin\beta$

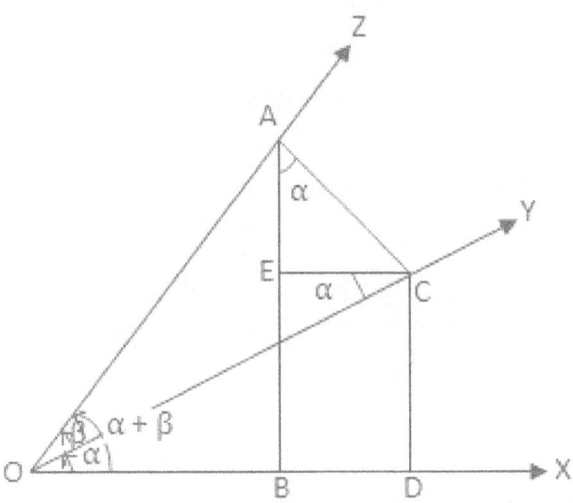

We will use the subsection of length $(OB = OD - EC)$ in the ratio of cos,

$$\cos(\alpha + \beta) = \frac{OB}{AO} = \frac{OD - EC}{AO}$$

Dividing and multiplying both the terms of numerator and denominator by OC and CA respectively, we have

$$= \frac{OD}{OC}\frac{OC}{AO} - \frac{EC}{CA}\frac{CA}{AO}$$

$$= \cos\alpha\cos\beta - \sin\alpha\sin\beta \qquad \text{QED}$$

Identity of difference of angles for cos:

Now we will prove the difference identity for cos. To do this, we use its sum identity together with negative identities. So difference identity is transformed into sum identity as follows:

$$\cos (\alpha - \beta) = \cos (\alpha + (-\beta))$$

Using sum identity of cos,

$$\cos (\alpha + (-\beta)) = \cos \alpha \cos (-\beta) - \sin \alpha \sin (-\beta)$$

From negative identity, $\cos (-\beta) = \cos \beta$, and $\sin (-\beta) = -\sin \beta$,

Hence,

$$\cos (\alpha - \beta) = \cos \alpha \cos \beta + \sin \alpha \sin \beta \qquad \text{QED}$$

Proof - Identities of Sum/Difference of Angles for Tan and Cot

Identity of sum of angles for tan:

Now we will prove the sum identity for tan.

$$\tan (\alpha + \beta) = \frac{\tan \alpha + \tan \beta}{1 - \tan \alpha \tan \beta} \text{ ; } \textit{where none of the angles } \alpha, \beta$$

$$\textit{and } \alpha + \beta \textit{ are an odd multiple of } \frac{\pi}{2}$$

We transform tan into sin & cos; and apply the identity of sum of angles.

$$\tan (\alpha + \beta) = \frac{\sin (\alpha + \beta)}{\cos (\alpha + \beta)}$$

For the given equation, angles α, β and $\alpha + \beta$ are not an odd multiple of $\dfrac{\pi}{2}$, so the denominator is not 0, hence the expression is valid.

Applying the sum formulas of sin at numerator and cos at denominator, we have

$$= \frac{\sin \alpha \cos \beta + \cos \alpha \sin \beta}{\cos \alpha \cos \beta - \sin \alpha \sin \beta}$$

Dividing numerator and denominator by cos α cos β, we have

$$= \frac{\dfrac{\sin \alpha \cos \beta}{\cos \alpha \cos \beta} + \dfrac{\cos \alpha \sin \beta}{\cos \alpha \cos \beta}}{\dfrac{\cos \alpha \cos \beta}{\cos \alpha \cos \beta} - \dfrac{\sin \alpha \sin \beta}{\cos \alpha \cos \beta}}$$

$$= \frac{\dfrac{\sin \alpha}{\cos \alpha} + \dfrac{\sin \beta}{\cos \beta}}{1 - \dfrac{\sin \alpha \sin \beta}{\cos \alpha \cos \beta}}$$

Hence,

$$\tan (\alpha + \beta) = \frac{\tan \alpha + \tan \beta}{1 - \tan \alpha \, \tan \beta} \qquad \text{QED.}$$

Identity of difference of angles for tan:

$$\tan (\alpha - \beta) = \frac{\tan \alpha - \tan \beta}{1 + \tan \alpha \, \tan \beta} \; ; \textit{ where none of the angles } \alpha, \beta$$
$$\textit{and } \alpha - \beta \textit{ are an odd multiple of } \frac{\pi}{2}$$

To verify the difference identity for tan, we use its sum identity besides using the appropriate sign. So difference formula is transformed into sum formula as follows:

$$\tan(\alpha - \beta) = \tan(\alpha + (-\beta))$$

Using sum identity of tan,

$$\tan(\alpha + (-\beta)) = \frac{\tan\alpha + \tan(-\beta)}{1 - \tan\alpha\,\tan(-\beta)}$$

Since $\tan(-\beta) = -\tan\beta$,

Hence,

$$\tan(\alpha - \beta) = \frac{\tan\alpha - \tan\beta}{1 + \tan\alpha\,\tan\beta} \qquad \text{QED}$$

Example-1: Find the value of $\tan 15°$ using identity of difference of angles for tan.

Solution: $\tan 15° = \tan(45° - 30°)$

Using the identity of difference of angles for tan, we have

$$\tan(45° - 30°) = \frac{\tan 45° - \tan 30°}{1 + \tan 45°\,\tan 30°}$$

$$= \frac{1 - \frac{1}{\sqrt{3}}}{1 + 1 \times \frac{1}{\sqrt{3}}} = \frac{\sqrt{3} - 1}{\sqrt{3} + 1}$$

Example-2: Find the value of $\tan\dfrac{13\pi}{12}$

Solution: $\tan\dfrac{13\pi}{12} = \tan\left(\pi + \dfrac{\pi}{12}\right) = \tan\dfrac{\pi}{12} = \tan\left(\dfrac{\pi}{4} - \dfrac{\pi}{6}\right)$

$$= \frac{\tan\frac{\pi}{4} - \tan\frac{\pi}{6}}{1 + \tan\frac{\pi}{4}\tan\frac{\pi}{6}} = \frac{1 - \frac{1}{\sqrt{3}}}{1 + \frac{1}{\sqrt{3}}} = \frac{\sqrt{3} - 1}{\sqrt{3} + 1} = 2 - \sqrt{3}$$

Identity of sum of angles for cot:

Now we will prove the sum identity for cot.

$$\cot(\alpha + \beta) = \frac{\cot\alpha\cot\beta - 1}{\cot\alpha + \cot\beta} \; ; \quad \textit{where none of the angles } \alpha, \beta$$

$$\textit{and } \alpha + \beta \textit{ are a multiple of } \pi$$

We transform cot sin and cos; and apply the corresponding identity of sum of angles.

$$\cot(\alpha + \beta) = \frac{\cos(\alpha + \beta)}{\sin(\alpha + \beta)}$$

For the given equation, any of angles α, β and $\alpha + \beta$ is not a multiple of $\frac{\pi}{2}$, otherwise the term becomes undefined.

Applying the sum formulas of cos at numerator and sin at denominator, we have

$$= \frac{\cos\alpha\cos\beta - \sin\alpha\sin\beta}{\sin\alpha\cos\beta + \cos\alpha\sin\beta}$$

Dividing numerator and denominator by $\sin\alpha\sin\beta$, we have

$$= \frac{\dfrac{\cos\alpha\cos\beta}{\sin\alpha\sin\beta} - \dfrac{\sin\alpha\sin\beta}{\sin\alpha\sin\beta}}{\dfrac{\sin\alpha\cos\beta}{\sin\alpha\sin\beta} + \dfrac{\cos\alpha\sin\beta}{\sin\alpha\sin\beta}}$$

$$= \frac{\dfrac{\cos \alpha \cos \beta}{\sin \alpha \sin \beta} - 1}{\dfrac{\cos \beta}{\sin \beta} + \dfrac{\cos \alpha}{\sin \alpha}}$$

Hence,

$$\cot (\alpha + \beta) = \frac{\cot \alpha \cot \beta - 1}{\cot \alpha + \cot \beta} \qquad \text{QED.}$$

Identity of difference of angles for cot:

$$\cot (\alpha - \beta) = \frac{\cot \alpha \cot \beta + 1}{\cot \beta - \cot \alpha} \; ; \quad \textit{where none of the angles } \alpha, \beta$$

$$\textit{and } \alpha - \beta \textit{ are a multiple of } \pi$$

To verify the difference identity for cot, we use its sum identity followed by negative identity. So difference formula is transformed into sum formula as follows:

$$\cot (\alpha - \beta) = \cot (\alpha + (- \beta))$$

Using sum identity of cot,

$$\cot (\alpha + (-\beta)) = \frac{\cot \alpha \cot (- \beta) - 1}{\cot \alpha + \cot (- \beta)}$$

From negative identity, $\cot (-\beta) = -\cot \beta$,

$$= \frac{-\cot \alpha \cot \beta - 1}{\cot \alpha - \cot \beta} = \frac{\cot \alpha \cot \beta + 1}{\cot \beta - \cot \alpha}$$

Hence,

$$\cot (\alpha - \beta) = \frac{\cot \alpha \cot \beta + 1}{\cot \beta - \cot \alpha} \qquad \text{QED.}$$

Identities of Sum/Difference of Angles:

$\sin(\alpha + \beta) = \sin\alpha\cos\beta + \cos\alpha\sin\beta$

$\sin(\alpha - \beta) = \sin\alpha\cos\beta - \cos\alpha\sin\beta$

$\cos(\alpha + \beta) = \cos\alpha\cos\beta - \sin\alpha\sin\beta$

$\cos(\alpha - \beta) = \cos\alpha\cos\beta + \sin\alpha\sin\beta$

$\tan(\alpha + \beta) = \dfrac{\tan\alpha + \tan\beta}{1 - \tan\alpha\tan\beta}$, *where none of the angles α, β and $(\alpha+\beta)$ are an odd multiple of $\dfrac{\pi}{2}$*

$\tan(\alpha - \beta) = \dfrac{\tan\alpha - \tan\beta}{1 + \tan\alpha\tan\beta}$, *where none of the angles α, β and $(\alpha-\beta)$ are an odd multiple of $\dfrac{\pi}{2}$*

$\cot(\alpha + \beta) = \dfrac{\cot\alpha\cot\beta - 1}{\cot\alpha + \cot\beta}$, *where none of the angles α, β and $(\alpha + \beta)$ are a multiple of π*

$\cot(\alpha - \beta) = \dfrac{\cot\alpha\cot\beta + 1}{\cot\beta - \cot\alpha}$, *where none of the angles α, β and $(\alpha - \beta)$ are a multiple of π*

Worksheet-10

Example-1: Find complementary angle of 70°

Example-2: Find complementary angle of −30°

Example-3: Find supplementary angle of 70°

Example-4: Find supplementary angle of −30°

Exercise-5: Find the value of $\tan \dfrac{13\,\pi}{12}$

Exercise-6: Find the value of tan 15° using difference identity.

Exercise-7: Find the value of cot 15° using difference identity.

Exercise-8: Find the value of sin 15° using difference identity.

Exercise-9: Find the value of sin 75° using difference identity.

Exercise-10: Find the value of cos 15° using difference identity.

Exercise-11: Find the value of cos 75° using difference identity.

Exercise-12: Verify the following formulas/identities:

a. $\dfrac{\sin(\alpha+\beta)}{\sin(\alpha-\beta)} = \dfrac{\tan\alpha+\tan\beta}{\tan\alpha-\tan\beta}$

b. $\tan(\alpha-\beta) = \dfrac{\tan\alpha-\tan\beta}{1+\tan\alpha\tan\beta}$

c. $\quad \cot(\alpha - \beta) = \dfrac{\cot\alpha\ \cot\beta\ + 1}{\cot\beta\ - \cot\alpha}$

d. $\quad \cot(\alpha + \beta) = \dfrac{\cot\alpha\ \cot\beta\ - 1}{\cot\alpha + \cot\beta}$

e. $\quad \tan(\alpha + \beta) = \dfrac{\tan\alpha + \tan\beta}{1 - \tan\alpha\ \tan\beta}$

Solutions-Worksheet-10

Solution-1: Complementary angle of $70° = 90°- 70° = 20°$

Solution-2: Complementary angle of $-30° = 90°- (-30°) = 120°$

Solution-3: Supplementary angle of $70° = 180°- 70° = 110°$

Solution-4: Supplementary angle of $-30° = 180°- (-30°) = 210°$

Solution-5:

$$\tan \frac{13\,\pi}{12} = \tan\left(\pi + \frac{\pi}{12}\right) = \tan \frac{\pi}{12} = \tan\left(\frac{\pi}{4} - \frac{\pi}{6}\right)$$

$$= \frac{\tan \frac{\pi}{4} - \tan \frac{\pi}{6}}{1 + \tan \frac{\pi}{4} \tan \frac{\pi}{6}} = \frac{1 - \frac{1}{\sqrt{3}}}{1 + \frac{1}{\sqrt{3}}} = \frac{\sqrt{3} - 1}{\sqrt{3} + 1}$$

$$= \frac{\sqrt{3} - 1}{\sqrt{3} + 1} \; \frac{\sqrt{3} - 1}{\sqrt{3} - 1} = 2 - \sqrt{3}$$

Solution-6: $\tan 15° = \tan(45° - 30°)$

$$\tan(45° - 30°) = \frac{\tan 45° - \tan 30°}{1 + \tan 45° \tan 30°}$$

$$= \frac{1 - \frac{1}{\sqrt{3}}}{1 + 1 \times \frac{1}{\sqrt{3}}} = \frac{\sqrt{3} - 1}{\sqrt{3} + 1}$$

$$= \frac{\sqrt{3} - 1}{\sqrt{3} + 1} \; \frac{\sqrt{3} - 1}{\sqrt{3} - 1} = 2 - \sqrt{3}$$

Solution-7: $\cot 15° = \cot(45° - 30°)$

$$\cot(45° - 30°) = \frac{\cot 45° \cot 30° + 1}{\cot 30° - \cot 45°}$$

$$= \frac{1 \times \sqrt{3} + 1}{\sqrt{3} - 1} = \frac{\sqrt{3} + 1}{\sqrt{3} - 1}$$

$$= \frac{\sqrt{3} + 1}{\sqrt{3} - 1} \frac{\sqrt{3} + 1}{\sqrt{3} + 1} = 2 + \sqrt{3}$$

Solution-8: $\sin 15° = \sin (45° - 30°)$

$$= \sin 45° \cos 30° - \cos 45° \sin 30°$$

$$= \frac{1}{\sqrt{2}} \frac{\sqrt{3}}{2} - \frac{1}{\sqrt{2}} \frac{1}{2}$$

$$= \frac{\sqrt{3}}{2\sqrt{2}} - \frac{1}{2\sqrt{2}} = \frac{\sqrt{3} - 1}{2\sqrt{2}}$$

Solution-9: $\sin 75° = \sin (45° + 30°)$

$$= \sin 45° \cos 30° + \cos 45° \sin 30°$$

$$= \frac{1}{\sqrt{2}} \frac{\sqrt{3}}{2} + \frac{1}{\sqrt{2}} \frac{1}{2}$$

$$= \frac{\sqrt{3}}{2\sqrt{2}} + \frac{1}{2\sqrt{2}} = \frac{\sqrt{3} + 1}{2\sqrt{2}}$$

Solution-10: $\cos 15° = \cos (45° - 30°)$

$$= \cos 45° \cos 30° + \sin 45° \sin 30°$$

$$= \frac{1}{\sqrt{2}} \frac{\sqrt{3}}{2} + \frac{1}{\sqrt{2}} \frac{1}{2}$$

$$= \frac{\sqrt{3}}{2\sqrt{2}} + \frac{1}{2\sqrt{2}} = \frac{\sqrt{3} + 1}{2\sqrt{2}}$$

Solution-11: $\cos 75° = \cos (45° + 30°)$

$$= \cos 45° \cos 30° - \sin 45° \sin 30°$$

$$= \frac{1}{\sqrt{2}} \frac{\sqrt{3}}{2} - \frac{1}{\sqrt{2}} \frac{1}{2}$$

$$= \frac{\sqrt{3}}{2\sqrt{2}} - \frac{1}{2\sqrt{2}} = \frac{\sqrt{3}-1}{2\sqrt{2}}$$

Solution-12-a:

$$\frac{\sin (\alpha + \beta)}{\sin (\alpha - \beta)} = \frac{\sin \alpha \cos \beta + \cos \alpha \sin \beta}{\sin \alpha \cos \beta - \cos \alpha \sin \beta}$$

Dividing numerator and denominator by $\cos \alpha \cos \beta$, we have

$$= \frac{\dfrac{\sin \alpha \cos \beta}{\cos \alpha \cos \beta} + \dfrac{\cos \alpha \sin \beta}{\cos \alpha \cos \beta}}{\dfrac{\sin \alpha \cos \beta}{\cos \alpha \cos \beta} - \dfrac{\cos \alpha \sin \beta}{\cos \alpha \cos \beta}}$$

$$= \frac{\tan \alpha + \tan \beta}{\tan \alpha - \tan \beta}$$

Solution-12-b-e: Solutions to be found at identities of sum/difference of angles explained in the section 4.2.

4.3   Double Angle Trigonometric Identities

Double angle identities are used to simplify the trigonometric expression. We will prove the double angle formulas using corresponding identities of sum of angles.

Double angle identities for sin:

$$\sin 2\alpha = 2 \sin \alpha \cos \alpha$$

Proof: We separate the double angle into two angles and apply the sum of angles identity for sin. So,

$$\sin 2\alpha = \sin (\alpha + \alpha) = \sin \alpha \cos \alpha + \cos \alpha \sin \alpha$$

$$= 2 \sin \alpha \cos \alpha \qquad \text{QED.}$$

Double angle formula for sin into tan

$$\sin 2\alpha = \frac{2 \tan \alpha}{1 + \tan^2 \alpha}$$

Proof: Double angle of sin transformed into sin and cos are:

$$\sin 2\alpha = 2 \sin \alpha \cos \alpha$$

Multiplying and dividing with cos α, we have

$$= 2 \sin \alpha \cos \alpha \; \frac{\cos \alpha}{\cos \alpha}$$

$$= 2 \left(\frac{\sin \alpha}{\cos \alpha} \right) \cos^2 \alpha$$

$$= 2 \tan \alpha \; \cos^2 \alpha$$

$$= 2 \tan \alpha \left(\frac{1}{\sec^2 \alpha} \right)$$

$$= \frac{2 \tan \alpha}{1 + \tan^2 \alpha} \text{ , where } \alpha \neq n\pi + \frac{\pi}{2}, \text{ for n as an integer } n \in \mathbb{Z}$$

Remark: if $\alpha = n\pi + \dfrac{\pi}{2}$ *; then the expression is undefined.*

Double angle identities for cos:

$$\cos 2\alpha = \cos^2 \alpha - \sin^2 \alpha$$

$$\cos 2\alpha = 1 - 2 \sin^2 \alpha$$

$$\cos 2\alpha = 2 \cos^2 \alpha - 1$$

Double angle formula transforming cos into square of sin and cos

Proof: We separate the double angle into two angles and apply the sum of angles identity of cos. Therefore,

$$\cos 2\alpha = \cos (\alpha + \alpha) = \cos \alpha \cos \alpha - \sin \alpha \sin \alpha$$

$$= \cos^2 \alpha - \sin^2 \alpha \qquad \text{QED}$$

Double angle formula transforming cos into square of sin

Proof: Using Pythagorean identity $\cos^2 \alpha + \sin^2 \alpha = 1$, we get other form of double angle.

Substituting for $\cos^2 \alpha$, we have

$$\cos 2\alpha = \cos^2 \alpha - \sin^2 \alpha = (1 - \sin^2 \alpha) - \sin^2 \alpha = 1 - 2\sin^2 \alpha$$

Double angle formula transforming cos into square of cos

Proof: Substituting in above equation for $\sin^2 \alpha$, we have

$$\cos 2\alpha = \cos^2 \alpha - \sin^2 \alpha = \cos^2 \alpha - (1 - \cos^2 \alpha) = 2\cos^2 \alpha - 1$$

Double angle formula for cos into tan

$$\cos 2\alpha = \frac{1 - \tan^2 \alpha}{1 + \tan^2 \alpha}$$

Proof: The double angle identity for cos is,

$$\cos 2\alpha = \cos^2 \alpha - \sin^2 \alpha$$

Multiplying and dividing with $\sec^2 \alpha$ to left hand side of equation, we have

$$= (\cos^2 \alpha - \sin^2 \alpha)\left(\frac{\sec^2 \alpha}{\sec^2 \alpha}\right)$$

$$= \frac{1 - \tan^2 \alpha}{\sec^2 \alpha}$$

Hence,

$$\cos 2\alpha = \frac{1 - \tan^2 \alpha}{1 + \tan^2 \alpha}$$

Double angle identities for tan:

$$\tan 2\alpha = \frac{2 \tan \alpha}{1 - \tan^2 \alpha}$$

Proof: We separate the double angle into two angles and apply the sum of angles identity for tan. Therefore,

$$\tan(\alpha + \alpha) = \frac{\tan \alpha + \tan \alpha}{1 - \tan \alpha \tan \alpha} = \frac{2 \tan \alpha}{1 - \tan^2 \alpha} \qquad \text{QED}$$

Double angle identities for cot:

$$\cot 2\alpha = \frac{\cot^2 \alpha - 1}{2 \cot \alpha}$$

Proof: We separate the double angle into two angles and apply the sum of angles identity for cot. Therefore,

$$\cot(\alpha + \alpha) = \frac{\cot \alpha \cot \alpha - 1}{\cot \alpha + \cot \alpha} = \frac{\cot^2 \alpha - 1}{2 \cot \alpha} \qquad \text{QED}$$

Double Angle Identities:

$$\sin 2\alpha = 2 \sin \alpha \cos \alpha$$

$$\sin 2\alpha = \frac{2 \tan \alpha}{1 + \tan^2 \alpha}$$

$$\cos 2\alpha = \cos^2 \alpha - \sin^2 \alpha$$

$$\cos 2\alpha = 1 - 2 \sin^2 \alpha$$

$$\cos 2\alpha = 2 \cos^2 \alpha - 1$$

$$\cos 2\alpha = \frac{1 - \tan^2 \alpha}{1 + \tan^2 \alpha}$$

$$\tan 2\alpha = \frac{2 \tan \alpha}{1 - \tan^2 \alpha}$$

$$\cot 2\alpha = \frac{\cot^2 \alpha - 1}{2 \cot \alpha}$$

Worksheet-11

Exercise-1: Verify the following.

$$\tan \alpha = \frac{1 - \cos 2\alpha}{\sin 2\alpha}$$

Exercise-2: Verify the following.

$$\tan \alpha = \frac{\sin 2\alpha}{1 + \cos 2\alpha}$$

Exercise-3: Verify the following.

$$\frac{1 - \cos 2\alpha}{\sin 2\alpha} = \frac{\sin 2\alpha}{1 + \cos 2\alpha}$$

Exercise-4: Apply sum of angles identity for cot and verify double angle formula for

$$\cot 2\alpha = \frac{\cot^2 \alpha - 1}{2 \cot \alpha}$$

What should be the condition so that equation is not undefined?

Exercise-5: Verify the following

$$\sin \alpha = \frac{2 \tan \frac{\alpha}{2}}{1 + \tan^2 \frac{\alpha}{2}}$$

Solutions-Worksheet-11

Solution-1: From the double angle identities, we have

$$1 - \cos 2\alpha = 2 \sin^2 \alpha \text{ ; and } \sin 2\alpha = 2 \sin \alpha \cos \alpha$$

Therefore,

$$\frac{1 - \cos 2\alpha}{\sin 2\alpha} = \frac{2 \sin^2 \alpha}{2 \sin \alpha \cos \alpha} = \tan \alpha \qquad \text{QED.}$$

Solution-2: From the double angle identities, we have

$$\sin 2\alpha = \frac{2 \tan \alpha}{1+\tan^2 \alpha} \text{ and } 1 + \cos 2\alpha = 2 \cos^2 \alpha$$

Therefore, we have

$$= \frac{\sin 2\alpha}{1+ \cos 2\alpha} = \frac{2 \tan \alpha}{1+\tan^2 \alpha} \frac{1}{2 \cos^2 \alpha}$$

$$= \frac{2 \tan \alpha}{\sec^2 \alpha} \frac{1}{2 \cos^2 \alpha} = \tan \alpha \qquad \text{QED.}$$

Solution-3: Using the double angle identities for sin and cos

$$\cos 2\alpha = 1 - 2 \sin^2 \alpha$$

$$\sin 2\alpha = 2 \sin \alpha \cos \alpha$$

Substituting at LHS of equation, we get

$$\frac{1 - \cos 2\alpha}{\sin 2\alpha} = \frac{2 \sin^2 \alpha}{2 \sin \alpha \cos \alpha} = \frac{2 \sin \alpha}{2 \cos \alpha}$$

$$= \frac{2 \sin \alpha}{2 \cos \alpha} \frac{\cos \alpha}{\cos \alpha} = \frac{\sin 2\alpha}{2 \cos^2 \alpha}$$

Substituting the double angle identity; $\cos 2\alpha = 2 \cos^2 \alpha - 1$, we get

$$= \frac{\sin 2\alpha}{2 \cos^2 \alpha} = \frac{\sin 2\alpha}{1 + \cos 2\alpha} \qquad \text{QED.}$$

Solution-4:

$$\cot 2\alpha = \cot (\alpha + \alpha) = \frac{\cot \alpha \cot \alpha - 1}{\cot \alpha + \cot \alpha} = \frac{\cot^2 \alpha - 1}{2 \cot \alpha}$$

Where angles α and 2α are not multiple of π.

Solution-5: $\sin \alpha = \sin 2 \left(\dfrac{\alpha}{2}\right)$

Using the double angle of sin transformed into sin and cos are:

$$\sin 2 \left(\frac{\alpha}{2}\right) = 2 \sin \frac{\alpha}{2} \cos \frac{\alpha}{2}$$

Multiplying and dividing with $\cos \dfrac{\alpha}{2}$, we have

$$2 \sin \frac{\alpha}{2} \cos \frac{\alpha}{2} = 2 \sin \frac{\alpha}{2} \cos \frac{\alpha}{2} \frac{\cos \frac{\alpha}{2}}{\cos \frac{\alpha}{2}} = 2 \left(\frac{\sin \frac{\alpha}{2}}{\cos \frac{\alpha}{2}}\right) \cos^2 \frac{\alpha}{2}$$

$$= 2 \tan \frac{\alpha}{2} \cos^2 \frac{\alpha}{2} = 2 \tan \frac{\alpha}{2} \left(\frac{1}{\sec^2 \frac{\alpha}{2}}\right)$$

$$= \frac{2 \tan \frac{\alpha}{2}}{1 + \tan^2 \frac{\alpha}{2}} \text{, where } \frac{\alpha}{2} \neq n\pi + \frac{\pi}{2}, n \in \mathbf{Z} \qquad \text{QED.}$$

4.4   Half Angle Trigonometric Identities

Half angle formulas are derived using corresponding sum of angles identities, specifically double angle identities.

Half angle formula for sin

$$\sin \frac{\alpha}{2} = \pm \sqrt{\frac{1- \cos \alpha}{2}}$$

As deduced before, one of the forms of double angle formula for cos is

$$\cos 2\alpha = 1 - 2 \sin^2 \alpha$$

If we reduce the angle α into half, we have

$$\cos \alpha = 1 - 2 \sin^2 \frac{\alpha}{2}$$

Rearranging $(\sin^2 \frac{\alpha}{2})$ to left hand side, we have

$$\sin^2 \frac{\alpha}{2} = \frac{1- \cos \alpha}{2}$$

Therefore,

$$\sin \frac{\alpha}{2} = \pm \sqrt{\frac{1- \cos \alpha}{2}}$$

The plus or minus sign will depend on the quadrant, as discussed before. For example, all trigonometric functions are positive in the first quadrant, i.e. for acute angles.

Half angle formula for cos

$$\cos\frac{\alpha}{2} = \pm\sqrt{\frac{1+\cos\alpha}{2}}$$

As deduced before, one of the forms of double angle formula for cos is

$$\cos 2\alpha = 2\cos^2\alpha - 1$$

If we reduce the angle α into half, we have

$$\cos\alpha = 2\cos^2\frac{\alpha}{2} - 1$$

Rearranging $(\cos^2\frac{\alpha}{2})$ to left hand side, we have

$$\cos^2\frac{\alpha}{2} = \frac{1+\cos\alpha}{2}$$

Therefore,

$$\cos\frac{\alpha}{2} = \pm\sqrt{\frac{1+\cos\alpha}{2}}$$

The plus or minus sign will depend on the quadrant, as discussed before. For instance, all trigonometric functions are positive in the first quadrant, i.e. for acute angles.

Half angle formula for tan

$$\tan\frac{\alpha}{2} = \frac{1-\cos\alpha}{\sin\alpha} = \frac{\sin\alpha}{1+\cos\alpha}$$

We use the ratio identity for tan and apply the half angle formula for sin and cos.

$$\tan \frac{\alpha}{2} = \frac{\sin \frac{\alpha}{2}}{\cos \frac{\alpha}{2}} = \frac{\sqrt{\dfrac{1-\cos\alpha}{2}}}{\sqrt{\dfrac{1+\cos\alpha}{2}}} = \sqrt{\frac{1-\cos\alpha}{1+\cos\alpha}}$$

Multiplying and diving the term by $\sqrt{1+\cos\alpha}$, we have

$$= \sqrt{\frac{1-\cos\alpha}{1+\cos\alpha}}\sqrt{\frac{1+\cos\alpha}{1+\cos\alpha}}$$

$$= \sqrt{\frac{1-\cos^2\alpha}{(1+\cos\alpha)^2}} = \sqrt{\frac{\sin^2\alpha}{(1+\cos\alpha)^2}}$$

Hence, $\tan \dfrac{\alpha}{2} = \dfrac{\sin\alpha}{1+\cos\alpha}$ QED

Besides this, multiplying and diving the term $\sqrt{\dfrac{1-\cos\alpha}{1+\cos\alpha}}$ by $\sqrt{1-\cos\alpha}$ leads to another identity,

$$\tan \frac{\alpha}{2} = \frac{1-\cos\alpha}{\sin\alpha}$$ QED

Half Angle Formulas:

$$\sin \frac{\alpha}{2} = \pm\sqrt{\frac{1-\cos\alpha}{2}}$$

$$\cos \frac{\alpha}{2} = \pm\sqrt{\frac{1+\cos\alpha}{2}}$$

$$\tan \frac{\alpha}{2} = \frac{\sin\alpha}{1+\cos\alpha} = \frac{1-\cos\alpha}{\sin\alpha}$$

4.5 Multiple Angle Formulas

In multi-angle transformation, we apply the corresponding double angle identity to transform to a simpler form.

$$\begin{aligned}
\sin 3\alpha \quad &= \sin(2\alpha + \alpha) \\
&= \sin 2\alpha \cos\alpha + \cos 2\alpha \sin\alpha \\
&= 2\sin\alpha\cos\alpha\cos\alpha + (\cos^2\alpha - \sin^2\alpha)\sin\alpha \\
&= 2\sin\alpha\cos^2\alpha + \sin\alpha\cos^2\alpha - \sin^3\alpha \\
&= 3\sin\alpha\cos^2\alpha - \sin^3\alpha
\end{aligned}$$

$$\begin{aligned}
\cos 3\alpha \quad &= \cos(2\alpha + \alpha) \\
&= \cos 2\alpha\cos\alpha - \sin 2\alpha\sin\alpha \\
&= (\cos^2\alpha - \sin^2\alpha)\cos\alpha - (2\sin\alpha\cos\alpha)\sin\alpha \\
&= \cos^3\alpha - \sin^2\alpha\cos\alpha - 2\sin^2\alpha\cos\alpha \\
&= \cos^3\alpha - 3\sin^2\alpha\cos\alpha
\end{aligned}$$

$$\begin{aligned}
\tan 3\alpha \quad &= \tan(2\alpha + \alpha) \\[2mm]
&= \frac{\tan 2\alpha + \tan\alpha}{1 - \tan 2\alpha\tan\alpha} \\[2mm]
&= \frac{\tan(\alpha+\alpha) + \tan\alpha}{1 - \tan(\alpha+\alpha)\tan\alpha} \\[2mm]
&= \frac{\dfrac{2\tan\alpha}{1 - \tan^2\alpha} + \tan\alpha}{1 - \dfrac{2\tan\alpha}{1 - \tan^2\alpha}\tan\alpha} \\[2mm]
&= \frac{2\tan\alpha + \tan\alpha - \tan^3\alpha}{1 - \tan^2\alpha - 2\tan^2\alpha} \\[2mm]
&= \frac{3\tan\alpha - \tan^3\alpha}{1 - 3\tan^2\alpha}
\end{aligned}$$

Similarly, we can find the value of sin 4α by applying the double angle identities repeatedly.

$\sin 4\alpha = \sin (2\alpha + 2\alpha)$

$$= \sin 2\alpha \cos 2\alpha + \cos 2\alpha \sin 2\alpha$$

$$= \sin (\alpha + \alpha) \cos (\alpha + \alpha) + \cos (\alpha + \alpha) \sin (\alpha + \alpha)$$

$$= (2 \sin \alpha \cos \alpha) (\cos^2 \alpha - \sin^2 \alpha)$$
$$+ (\cos^2 \alpha - \sin^2 \alpha) (2 \sin \alpha \cos \alpha)$$

$$= (2 \sin \alpha \cos^3 \alpha - 2 \cos \alpha \sin^3 \alpha)$$
$$+ (2 \sin \alpha \cos^3 \alpha - 2 \cos \alpha \sin^3 \alpha)$$

$$= 4 \sin \alpha \cos^3 \alpha - 4 \cos \alpha \sin^3 \alpha$$

$$= 2 (2 \sin \alpha \cos \alpha) (\cos^2 \alpha - \sin^2 \alpha)$$

$$= 2 \sin 2\alpha \cos 2\alpha$$

Multiple Angle Formulas:

$$\sin 3\alpha = = 3 \sin \alpha \cos^2 \alpha - \sin^3 \alpha$$

$$\cos 3\alpha = \cos^3 \alpha - 3 \sin^2 \alpha \cos \alpha$$

$$\tan 3\alpha = \frac{3 \tan \alpha - \tan^3 \alpha}{1 - 3 \tan^2 \alpha}$$

$$\sin 4\alpha = 2 \sin 2\alpha \cos 2\alpha$$

Worksheet-12

Exercise-1: Verify the following

$$\tan\frac{\alpha}{2} = \frac{1-\cos\alpha}{\sin\alpha}$$

Exercise-2: Verify the following

$$\tan\frac{\alpha}{2} = \frac{\tan\alpha}{\sec\alpha+1}$$

Exercise-3: Evaluate $\cos\frac{\pi}{8}$

Exercise-4: Show that

$$2 \sin \frac{\pi}{6} \ \sec \frac{\pi}{3} - 2 \sin \frac{5\pi}{6} \cot \frac{\pi}{4} = 1$$

Exercise-5: Prove that

$$\cos 3\alpha = 4 \ \cos^3 \alpha - 3 \cos \alpha$$

Exercise-6: Prove that

$$\sin 3\alpha = 3 \sin \alpha - 4 \sin^3 \alpha$$

Exercise-7: Verify

$$\cos 4\alpha = \cos^4 \alpha + \sin^4 \alpha - 6 \cos^2\alpha \sin^2 \alpha$$

Exercise-8: Verify

$$\tan 4\alpha = \frac{4 \tan \alpha \, (1 - \tan^2 \alpha)}{1 - 6 \tan^2 \alpha + \tan^4 \alpha}$$

Solutions-Worksheet-12

Soution-1: Refer the half angle formula for tan in the section 4.4.

$$\tan\frac{\alpha}{2} = \frac{1 - \cos\alpha}{\sin\alpha}$$

Solution-2: The half angle formula for is,

$$\tan\frac{\alpha}{2} = \frac{\sin\alpha}{1+\cos\alpha}$$

Dividing the numerator and denominator by cos at right hand side (RHS) of equation, we get $\dfrac{\tan\alpha}{\sec\alpha+1}$ QED

Solution-3: Angle $\dfrac{\pi}{8}$ is half of $\dfrac{\pi}{4}$, so we apply half angle formula for cos, where we consider only positive value because angle $\dfrac{\pi}{8}$ is an acute angle, i.e. in first quadrant.

$$\cos\frac{\pi}{8} = \sqrt{\frac{1+\cos\frac{\pi}{4}}{2}}$$

$$= \sqrt{\frac{1+\frac{1}{\sqrt{2}}}{2}} = \sqrt{\frac{\frac{\sqrt{2}+1}{\sqrt{2}}}{2}} = \sqrt{\frac{\sqrt{2}+1}{2\sqrt{2}}}$$

Multiplying numerator and denominator by $\sqrt{2}$, we get

$$= \frac{1}{2}\sqrt{(2+\sqrt{2})}$$

Solution-4:

$$2\sin\frac{\pi}{6}\sec\frac{\pi}{3} - 2\sin\frac{5\pi}{6}\cot\frac{\pi}{4}$$

$$= (2 \times \frac{1}{2} \times 2) - (2 \times \sin(\pi - \frac{\pi}{6}) \times 1) = 2 - 2 \sin \frac{\pi}{6}$$

$$= 2 - 2 \times \frac{1}{2} = 1$$

Solution-5: We use the double angle formula and transform to simpler form.

$$\cos 3\alpha \quad = \cos(2\alpha + \alpha) = \cos^3 \alpha - 3 \sin^2 \alpha \cos \alpha$$

Replacing $\sin^2 \alpha$ in RHS with $(1 - \cos^2 \alpha)$, we have

$$= \cos^3 \alpha - 3(1 - \cos^2 \alpha) \cos \alpha$$

Therefore, $\cos 3\alpha = 4 \cos^3 \alpha - 3 \cos \alpha$; hence proved.

Solution-6: We use the double angle formula and transform to simpler form.

$$\sin 3\alpha \quad = 3 \sin \alpha \cos^2 \alpha - \sin^3 \alpha$$

Replacing $\cos^2 \alpha$ in RHS with $(1 - \sin^2 \alpha)$, we have

$$= 3 \sin \alpha (1 - \sin^2 \alpha) - \sin^3 \alpha$$

Therefore, $\sin 3\alpha = 3 \sin \alpha - 4 \sin^3 \alpha$ \qquad QED

Solution-7: Simplifying using double angle,

$$\cos 4\alpha = \cos(2\alpha + 2\alpha)$$

$$= \cos 2\alpha \cos 2\alpha - \sin 2\alpha \sin 2\alpha$$

Substituting $\cos 2\alpha = \cos^2 \alpha - \sin^2 \alpha$; and $\sin 2\alpha = 2 \sin \alpha \cos \alpha$, we get

$$= (\cos^2 \alpha - \sin^2 \alpha)(\cos^2 \alpha - \sin^2 \alpha)$$

$$- (2 \sin \alpha \cos \alpha)(2 \sin \alpha \cos \alpha)$$

$$= (\cos^4 \alpha + \sin^4 \alpha - 2\cos^2\alpha \sin^2 \alpha)$$

$$- (4\sin^2\alpha \cos^2 \alpha)$$

$$= \cos^4 \alpha + \sin^4 \alpha - 6\cos^2\alpha \sin^2 \alpha \qquad \text{QED.}$$

Solution-8: Simplifying using double angle,

$$\tan 4\alpha = \frac{2\tan 2\alpha}{1 - \tan^2 2\alpha}$$

$$= \frac{\dfrac{4\tan\alpha}{1 - \tan^2 \alpha}}{1 - \dfrac{4\tan^2 \alpha}{(1 - \tan^2 \alpha)^2}}$$

$$= \frac{\dfrac{4\tan\alpha}{1 - \tan^2 \alpha}}{\dfrac{1 - 2\tan^2 \alpha + \tan^4 \alpha - 4\tan^2 \alpha}{(1 - \tan^2 \alpha)^2}}$$

$$= \frac{4\tan\alpha\,(1 - \tan^2 \alpha)}{1 - 6\tan^2 \alpha + \tan^4 \alpha} \qquad ; \text{Q.E.D}$$

4.6   Sum as Product Identities

Identities of sum as product are simple to verify. We use the trigonometric identities for sum/difference of angles to prove these formulas.

<u>**Sum as product for sin**</u>

$\sin(\alpha + \beta) + \sin(\alpha - \beta) = 2\sin\alpha\cos\beta$

$\sin(\alpha + \beta) - \sin(\alpha - \beta) = 2\cos\alpha\sin\beta$

Alternatively, Let the angle $\alpha + \beta = a$, and $\alpha - \beta = b$, then

$$\sin a + \sin b = 2\sin\frac{a+b}{2}\cos\frac{a-b}{2}$$

$$\sin a - \sin b = 2\cos\frac{a+b}{2}\sin\frac{a-b}{2}$$

<u>**Proof**</u>: As discussed earlier, trigonometric identities involving sum/difference of angles for sin are:

$\sin(\alpha + \beta) = \sin\alpha\cos\beta + \cos\alpha\sin\beta$ --------Eq. (1)

$\sin(\alpha - \beta) = \sin\alpha\cos\beta - \cos\alpha\sin\beta$ --------Eq. (2)

Adding the equation (1) and equation (2), we have

$\sin(\alpha + \beta) + \sin(\alpha - \beta) = 2\sin\alpha\cos\beta$ QED.

Subtracting equation (2) from equation (1), we have

$$\sin(\alpha + \beta) - \sin(\alpha - \beta) = 2\cos\alpha\sin\beta \qquad \text{QED.}$$

Let the angle $\alpha + \beta = a$, and $\alpha - \beta = b$, therefore

$$\alpha = \frac{a+b}{2} \text{ and } \beta = \frac{a-b}{2}$$

Substituting for angles, sum as product for sin can be formulated as:

$$\sin a + \sin b = 2\sin\frac{a+b}{2}\cos\frac{a-b}{2} \qquad \text{QED.}$$

$$\sin a - \sin b = 2\cos\frac{a+b}{2}\sin\frac{a-b}{2} \qquad \text{QED.}$$

Sum as Product for cos

$$\cos(\alpha + \beta) + \cos(\alpha - \beta) = 2\cos\alpha\cos\beta$$

$$\cos(\alpha + \beta) - \cos(\alpha - \beta) = -2\sin\alpha\sin\beta$$

Alternatively, let the angle $\alpha + \beta = a$, and $\alpha - \beta = b$, then

$$\cos a + \cos b = 2\cos\frac{a+b}{2}\cos\frac{a-b}{2}$$

$$\cos a - \cos b = -2\sin\frac{a+b}{2}\sin\frac{a-b}{2}$$

Proof: As discussed earlier, trigonometric identities involving sum/difference of angles for cos are:

$$\cos (\alpha + \beta) = \cos \alpha \cos \beta - \sin \alpha \sin \beta \qquad \text{-------Eq. (3)}$$

$$\cos (\alpha - \beta) = \cos \alpha \cos \beta + \sin \alpha \sin \beta \qquad \text{-------Eq. (4)}$$

Adding the equation (3) and equation (4), we have

$$\cos (\alpha + \beta) + \cos (\alpha - \beta) = 2 \cos \alpha \cos \beta \qquad \text{QED.}$$

Subtracting equation (4) from equation (3), we have

$$\cos (\alpha + \beta) - \cos (\alpha - \beta) = -2 \sin \alpha \sin \beta \qquad \text{QED.}$$

Let the angle $\alpha + \beta = a$, and $\alpha - \beta = b$, therefore

$$\alpha = \frac{a+b}{2} \text{ and } \beta = \frac{a-b}{2}$$

Substituting for angles, sum as product for cos can be formulated as:

$$\cos a + \cos b = 2 \cos \frac{a+b}{2} \cos \frac{a-b}{2} \qquad \text{QED.}$$

$$\cos a - \cos b = -2 \sin \frac{a+b}{2} \sin \frac{a-b}{2} \qquad \text{QED.}$$

Sum as Product Identities:

$$\sin(\alpha + \beta) + \sin(\alpha - \beta) = 2\sin\alpha\cos\beta$$

$$\sin(\alpha + \beta) - \sin(\alpha - \beta) = 2\cos\alpha\sin\beta$$

$$\cos(\alpha + \beta) + \cos(\alpha - \beta) = 2\cos\alpha\cos\beta$$

$$\cos(\alpha + \beta) - \cos(\alpha - \beta) = -2\sin\alpha\sin\beta$$

$$\sin a + \sin b = 2\sin\frac{a+b}{2}\cos\frac{a-b}{2}$$

$$\sin a - \sin b = 2\cos\frac{a+b}{2}\sin\frac{a-b}{2}$$

$$\cos a + \cos b = 2\cos\frac{a+b}{2}\cos\frac{a-b}{2}$$

$$\cos a - \cos b = -2\sin\frac{a+b}{2}\sin\frac{a-b}{2}$$

4.7   Product as Sum Identities

Identities of product as sum are just alternative or other form of sum as product. This involves interchanging the left side with right side of equations. Different product-as-sum identities are as follows:

$$2\sin\alpha\cos\beta = \sin(\alpha + \beta) + \sin(\alpha - \beta)$$
$$2\cos\alpha\sin\beta = \sin(\alpha + \beta) - \sin(\alpha - \beta)$$
$$2\cos\alpha\cos\beta = \cos(\alpha + \beta) + \cos(\alpha - \beta)$$
$$2\sin\alpha\sin\beta = -\cos(\alpha + \beta) + \cos(\alpha - \beta)$$

Let the angle $\alpha + \beta = a$, and $\alpha - \beta = b$, then

$$2 \sin \frac{a+b}{2} \cos \frac{a-b}{2} = \sin a + \sin b$$

$$2 \cos \frac{a+b}{2} \sin \frac{a-b}{2} = \sin a - \sin b$$

$$2 \cos \frac{a+b}{2} \cos \frac{a-b}{2} = \cos a + \cos b$$

$$2 \sin \frac{a+b}{2} \sin \frac{a-b}{2} = -\cos a + \cos b$$

Product as Sum Identities:

$$2 \sin \alpha \cos \beta = \sin(\alpha + \beta) + \sin(\alpha - \beta)$$
$$2 \cos \alpha \sin \beta = \sin(\alpha + \beta) - \sin(\alpha - \beta)$$
$$2 \cos \alpha \cos \beta = \cos(\alpha + \beta) + \cos(\alpha - \beta)$$
$$2 \sin \alpha \sin \beta = \cos(\alpha - \beta) - \cos(\alpha + \beta)$$

$$2 \sin \frac{a+b}{2} \cos \frac{a-b}{2} = \sin a + \sin b$$

$$2 \cos \frac{a+b}{2} \sin \frac{a-b}{2} = \sin a - \sin b$$

$$2 \cos \frac{a+b}{2} \cos \frac{a-b}{2} = \cos a + \cos b$$

$$2 \sin \frac{a+b}{2} \sin \frac{a-b}{2} = \cos b - \cos a$$

Worksheet-13

Exercise-1: Prove that $\dfrac{\cos 7\alpha + \cos 5\alpha}{\sin 7\alpha - \sin 5\alpha} = \cot \alpha$

Exercise-2: Prove that

$$\cot^2 \frac{\pi}{6} + \csc \frac{5\pi}{6} + \tan^2 \frac{\pi}{6} = \frac{16}{3}$$

Example-3: Verify

$$4 \sin^2 \frac{\pi}{6} + 4 \cos^2 \frac{\pi}{3} - \tan^2 \frac{\pi}{4} = 1$$

Example-4: Verify

$$4 \sin^2 \frac{\pi}{6} + \operatorname{cosec}^2 \frac{7\pi}{6} \cos^2 \frac{\pi}{3} = 2$$

Exercise-5: Prove that

$$\frac{\sin 5\alpha - 2\sin 3\alpha + \sin \alpha}{\cos 5\alpha - \cos \alpha} = \tan \alpha$$

Exercise-6: Verify

$$\cos\left(\frac{\pi}{4} - \alpha\right)\cos\left(\frac{\pi}{4} - \beta\right) - \sin\left(\frac{\pi}{4} - \alpha\right)\sin\left(\frac{\pi}{4} - \beta\right)$$

$$= \sin(\alpha + \beta)$$

Exercise-7: Prove that

$$\frac{\tan\left(\frac{\pi}{4} + \beta\right)}{\tan\left(\frac{\pi}{4} - \beta\right)} = \left(\frac{1 + \tan\beta}{1 - \tan\beta}\right)^2$$

Exercise-8: Prove that

$$\frac{\cos (\pi + \beta) \cos(-\beta)}{\sin (\pi - \beta) \cos (\frac{\pi}{2} + \beta)} = \cot^2 \beta$$

Exercise-9: Verify

$$\cos(\frac{3\pi}{2} + \beta) \cos(2\pi + \beta) [\cot(\frac{3\pi}{2} - \beta) + \cot(2\pi + \beta)] = 1$$

Exercise-10: Verify

$$\sin(n+1)\beta \sin(n+2)\beta + \cos(n+1)\beta \cos(n+2)\beta = \cos\beta$$

Exercise-11: Prove that

$$\cos(\frac{3\pi}{4} + \beta) - \cos(\frac{3\pi}{4} - \beta) = -\sqrt{2}\ \sin\beta$$

Exercise-12: Prove that

$$\sin^2 6\beta - \sin^2 4\beta = \sin 2\beta \sin 10\beta$$

Exercise-13: Verify

$$\cos^2 2\beta - \cos^2 6\beta = \sin 4\beta \ \sin 8\beta$$

Exercise-14: Verify

$$\sin 2\beta + 2 \sin 4\beta + \sin 6\beta = 4 \cos^2 \beta \sin 4\beta$$

Exercise-15: Prove that

$$\cot 4\beta \ (\sin 5\beta + \sin 3\beta) = \cot \beta \ (\sin 5\beta - \sin 3\beta\ )$$

Exercise-16: Prove that

$$\frac{\cos 9\beta - \cos 5\beta}{\sin 17\beta - \sin 3\beta} = -\frac{\sin 2\beta}{\cos 10\beta}$$

Exercise-17: Verify

$$\frac{\sin 5\beta + \sin 3\beta}{\cos 5\beta + \cos 3\beta} = \tan 4\beta$$

Exercise-18: Prove that

$$\frac{\sin \alpha - \sin \beta}{\cos \alpha + \cos \beta} = \tan \frac{\alpha - \beta}{2}$$

Exercise-19: Prove that

$$\frac{\sin \beta + \sin 3\beta}{\cos \beta + \cos 3\beta} = \tan 2\beta$$

Exercise-20: Prove that

$$\frac{\sin \beta - \sin 3\beta}{\sin^2 \beta - \cos^2 \beta} = 2 \sin \beta$$

Exercise-21: Verify

$$\tan 4\alpha = \frac{4\tan\alpha\,(1-\tan^2\alpha)}{1-6\tan^2\alpha+\tan^4\alpha}$$

Exercise-22: Prove that

$$\cos 4\beta = 1 - 8\sin^2\beta\cos^2\beta$$

Exercise-23: Prove that

$$\cot\beta\cot 2\beta - \cot 2\beta\cot 3\beta - \cot\beta\cot 3\beta = 1$$

Solutions-Worksheet-13

Solution-1: Simplifying the LHS, we have

$$= \frac{2 \cos \frac{7\alpha + 5\alpha}{2} \cos \frac{7\alpha - 5\alpha}{2}}{2 \cos \frac{7\alpha + 5\alpha}{2} \sin \frac{7\alpha - 5\alpha}{2}} = \frac{\cos \alpha}{\sin \alpha} = \cot \alpha \quad \text{Q.E.D}$$

Solution-2: In LHS, cosec is changed into sin, and supplementary identity of sin is used. So, $\sin \frac{5\pi}{6} = \sin(\pi - \frac{\pi}{6}) = \sin \frac{\pi}{6}$. Rearranging the LHS, we have

$$= \cot^2 \frac{\pi}{6} + \frac{1}{\sin(\frac{5\pi}{6})} + \tan^2 \frac{\pi}{6}$$

$$= \cot^2 \frac{\pi}{6} + \frac{1}{\sin \frac{\pi}{6}} + \tan^2 \frac{\pi}{6}$$

Placing the values, we have

$$= \left(\sqrt{3}\right)^2 + \frac{1}{(\frac{1}{2})} + \left(\frac{1}{\sqrt{3}}\right)^2$$

$$= \left(\sqrt{3}\right)^2 + \frac{2}{1} + \left(\frac{1}{\sqrt{3}}\right)^2$$

$$= 3 + 2 + \frac{1}{3} = \frac{16}{3} \quad \quad \text{Q.E.D}$$

Solution-3: Placing the values in LHS, we have

$$= 4 \left(\frac{1}{2}\right)^2 + 4 \left(\frac{1}{2}\right)^2 - 1 = 1 \quad \quad \text{Q.E.D}$$

Solution-4: In LHS, changing the cosec into sin, we have

$$= 4\sin^2\frac{\pi}{6} + \frac{1}{\sin^2(\pi + \frac{\pi}{6})}\cos^2\frac{\pi}{3}$$

We will apply sum/difference identity for sin. (*Alternatively, we can also use the sin* ($\pi + \theta$) *identity*)

$$= 4\sin^2\frac{\pi}{6} + (\frac{1}{\sin\pi\cos\frac{\pi}{6} + \cos\pi\sin\frac{\pi}{6}})^2\cos^2\frac{\pi}{3}$$

Placing the values, we have

$$= 4\sin^2\frac{\pi}{6} + (\frac{1}{(0)(\frac{\sqrt{3}}{2}) + (-1)(\frac{1}{2})})^2\cos^2\frac{\pi}{3}$$

$$= 4(\frac{1}{2})^2 + (-2)^2(\frac{1}{2})^2 = 1 + 1 = 2 \qquad \text{Q.E.D}$$

Solution-5: Working on LHS, we have

$$= \frac{\sin 5\alpha + \sin\alpha - 2\sin 3\alpha}{\cos 5\alpha - \cos\alpha}$$

$$= \frac{2\sin 3\alpha\cos 2\alpha - 2\sin 3\alpha}{-2\sin 3\alpha\sin 2\alpha} = \frac{-2\sin 3\alpha(\cos 2\alpha - 1)}{2\sin 3\alpha\sin 2\alpha}$$

$$= \frac{1 - \cos 2\alpha}{\sin 2\alpha} = \frac{2\sin^2\alpha}{2\sin\alpha\cos\alpha} = \tan\alpha \qquad \text{Q.E.D}$$

Solution-6: Let us consider, $\left(\frac{\pi}{4} - \alpha\right) = A$, and $\left(\frac{\pi}{4} - \beta\right) = B$. Substituting the angles A and B at LHS, we have

$$\cos A\cos B - \sin A\sin B$$

Applying the difference of angle identity for cos, we get

$$\cos A\cos B - \sin A\sin B = \cos(A + B)$$

So putting in the values for A and B, we get

$$\cos (A+B) = \cos \left(\frac{\pi}{4} - \alpha + \frac{\pi}{4} - \beta\right)$$

$$= \cos \left(\frac{\pi}{2} - (\alpha + \beta)\right) = \sin (\alpha + \beta) \qquad \text{Q.E.D}$$

Solution-7: Applying the sum and difference identities for tan in LHS, we have

$$= \frac{\dfrac{\tan \frac{\pi}{4} + \tan \beta}{1 - \tan \frac{\pi}{4} \tan \beta}}{\dfrac{\tan \frac{\pi}{4} - \tan \beta}{1 + \tan \frac{\pi}{4} \tan \beta}} = \frac{\dfrac{1 + \tan \beta}{1 - \tan \beta}}{\dfrac{1 - \tan \beta}{1 + \tan \beta}} = \left(\frac{1 + \tan \beta}{1 - \tan \beta}\right)^2 \qquad \text{Q.E.D}$$

Solution-8: We will use the identities of $(\pi+\beta)$, $(\pi-\beta)$, $\left(\frac{\pi}{2} + \beta\right)$ etc, as follows:

$$\cos (\pi+\beta) = - \cos \beta \; ; \; \cos (-\beta) = \cos \beta \; ; \; \sin (\pi-\beta) = \sin \beta;$$

$$\cos \left(\frac{\pi}{2} + \beta\right) = - \sin \beta \; ;$$

Applying the values in LHS of equation, we have

$$\frac{\cos (\pi + \beta) \cos(-\beta)}{\sin (\pi - \beta) \cos \left(\frac{\pi}{2} + \beta\right)} = \frac{- \cos \beta \cos \beta}{(\sin \beta) (- \sin \beta)}$$

$$= \frac{- \cos^2 \beta}{- \sin^2 \beta} = \cot^2 \beta \qquad \text{Q.E.D}$$

Solution-9: Simplifying the different terms of LHS,

$$\cot(2\pi + \beta) = \cot \beta$$

$$\cot\left(\frac{3\pi}{2} - \beta\right) = \cot\left(\pi + \frac{\pi}{2} - \beta\right) = \cot\left(\frac{\pi}{2} - \beta\right) = \tan \beta$$

$$\cos(2\pi + \beta) = \cos \beta$$

$$\cos\left(\frac{3\pi}{2} + \beta\right) = \cos\left(\pi + \frac{\pi}{2} + \beta\right) = -\cos\left(\frac{\pi}{2} + \beta\right) = \sin \beta$$

Substituting the simplified terms in LHS,

$$\cos(\frac{3\pi}{2} + \beta) \cos(2\pi + \beta) \, [\, \cot(\frac{3\pi}{2} - \beta) + \cot(2\pi + \beta) \,]$$

$$= \sin \beta \cos \beta \, [\, \tan \beta + \cot \beta \,]$$

$$= [\sin^2 \beta + \cos^2 \beta] = 1$$

Solution-10: Simplifying the first tem of LHS, we have

$$\sin (n + 1)\beta \sin (n + 2)\beta$$

$$= -\frac{1}{2} \cos (n + 2 + n + 1) \, \beta + \frac{1}{2} \cos (n + 2 - n - 1) \, \beta$$

$$= -\frac{1}{2} \cos (2n + 3) \, \beta + \frac{1}{2} \cos \beta$$

Simplifying the second tem of LHS, we have

$$\cos (n + 1)\beta \cos (n + 2)\beta$$

$$= \frac{1}{2} \cos (n + 2 + n + 1) \, \beta + \frac{1}{2} \cos (n + 2 - n - 1) \, \beta$$

$$= \frac{1}{2} \cos (2n + 3) \, \beta + \frac{1}{2} \cos \beta$$

Rearranging both the terms of LHS together, we have

$$-\frac{1}{2} \cos (2n + 3) \, \beta + \frac{1}{2} \cos \beta + \frac{1}{2} \cos (2n + 3) \, \beta + \frac{1}{2} \cos \beta$$

$$= \frac{1}{2} \, (\cos \beta + \cos \beta) = \cos \beta \qquad\qquad \text{Q.E.D}$$

Solution-11: Working on LHS of equation, we have

$$\cos(\frac{3\pi}{4} + \beta) - \cos(\frac{3\pi}{4} - \beta) = \cos(\pi - \frac{\pi}{4} + \beta) - \cos(\pi - \frac{\pi}{4} - \beta)$$

$$= \cos(\pi - (\tfrac{\pi}{4} - \beta)) - \cos(\pi - (\tfrac{\pi}{4} + \beta))$$

$$= -\cos(\tfrac{\pi}{4} - \beta) - (-\cos(\tfrac{\pi}{4} + \beta))$$

$$= \cos(\tfrac{\pi}{4} + \beta) - \cos(\tfrac{\pi}{4} - \beta)$$

$$= -2 \sin \tfrac{\pi}{4} \sin \beta = -2 \tfrac{1}{\sqrt{2}} \sin \beta = -\sqrt{2} \sin \beta$$

Hence proved.

Solution-12: Rearranging the term of LHS and applying the sum/difference identities for sin, we have

$$\sin^2 6\beta - \sin^2 4\beta = (\sin 6\beta + \sin 4\beta)(\sin 6\beta - \sin 4\beta)$$

$$= (2 \sin 5\beta \cos \beta)(2 \cos 5\beta \sin \beta))$$

$$= (2 \sin \beta \cos \beta)(2 \sin 5\beta \cos 5\beta) = \sin 2\beta \sin 10\beta; \text{ QED}.$$

Solution-13: Rearranging the term of LHS and applying the sum/difference identities for cos, we get

$$\cos^2 2\beta - \cos^2 6\beta = (\cos 2\beta + \cos 6\beta)(\cos 2\beta - \cos 6\beta)$$

$$= (2 \cos 4\beta \cos 2\beta)(-2 \sin 4\beta \sin(-2\beta))$$

$$= (2 \cos 4\beta \cos 2\beta)(2 \sin 4\beta \sin 2\beta)$$

$$= (2 \sin 2\beta \cos 2\beta)(2 \sin 4\beta \cos 4\beta)$$

$$= \sin 4\beta \sin 8\beta \qquad \text{Q.E.D}$$

Solution-14: Rearranging the LHS and applying the sum identity for $(\sin 6\beta + \sin 2\beta)$, we have,

$$\sin 6\beta + \sin 2\beta + 2 \sin 4\beta = 2 \sin 4\beta \cos 2\beta + 2 \sin 4\beta$$

$$= 2 \sin 4\beta (\cos 2\beta + 1) = 2 \sin 4\beta (2 \cos^2 \beta) = 4 \cos^2 \beta \sin 4\beta$$

Q.E.D

Solution-15: Taking the term cot 4β to RHS and (sin 5β – sin 3β) to LHS, we have

$$\frac{\sin 5\beta + \sin 3\beta}{\sin 5\beta - \sin 3\beta} = \frac{\cot \beta}{\cot 4\beta}$$

Expanding LHS using sum/difference identities, we have,

$$\frac{\sin 5\beta + \sin 3\beta}{\sin 5\beta - \sin 3\beta} = \frac{2 \sin 4\beta \cos \beta}{2 \cos 4\beta \sin \beta} = \frac{\cot \beta}{\cot 4\beta} \quad \text{Hence proved.}$$

Solution-16: Applying the sum identities for sin and cos in LHS, we have,

$$\frac{\cos 9\beta - \cos 5\beta}{\sin 17\beta - \sin 3\beta} = \frac{-2 \sin 7\beta \sin 2\beta}{2 \cos 10\beta \sin 7\beta} = -\frac{\sin 2\beta}{\cos 10\beta} \quad \text{Q.E.D}$$

Solution-17: Applying the sum identities for sin and cos in LHS, we have,

$$\frac{\sin 5\beta + \sin 3\beta}{\cos 5\beta + \cos 3\beta} = \frac{2 \sin 4\beta \cos \beta}{2 \cos 4\beta \cos \beta} = \frac{2 \sin 4\beta}{2 \cos 4\beta} = \tan 4\beta \quad \text{Q.E.D}$$

Solution-18: Applying the sum identity for sin and cos in LHS, we have,

$$\frac{\sin \alpha - \sin \beta}{\cos \alpha + \cos \beta} = \frac{2 \cos \frac{\alpha+\beta}{2} \sin \frac{\alpha-\beta}{2}}{2 \cos \frac{\alpha+\beta}{2} \cos \frac{\alpha-\beta}{2}} = \tan \frac{\alpha - \beta}{2} \quad \text{Q.E.D}$$

Solution-19: Applying the sum identities for sin and cos in LHS, we have,

$$\frac{\sin \beta + \sin 3\beta}{\cos \beta + \cos 3\beta} = \frac{2 \sin 2\beta \cos \beta}{2 \cos 2\beta \cos \beta} = \frac{2 \sin 2\beta}{2 \cos 2\beta} = \tan 2\beta \quad \text{Q.E.D}$$

Solution-20: Multiplying numerator and denominators of LHS by

−1, we have

$$\frac{\sin 3\beta - \sin \beta}{\cos^2\beta - \sin^2\beta}$$

The terms are simplified using the sum as product identity of sin; and double angle identity of cos,

$$\sin 3\beta - \sin \beta = 2 \cos 2\beta \ \sin \beta$$

$$\cos^2\beta - \sin^2\beta = \cos 2\beta$$

Substituting the simplified terms in LHS, we get

$$\frac{\sin 3\beta - \sin \beta}{\cos^2\beta - \sin^2\beta} = \frac{2 \cos 2\beta \sin \beta}{\cos 2\beta} = 2 \sin \beta \qquad \text{Q.E.D}$$

Solution-21: Applying the double angle identity in LHS, we have

$$\tan 4\alpha = \frac{2 \tan 2\alpha}{1 - \tan^2 2\alpha}$$

Applying the double angle identity for tan again, we get

$$= \frac{\dfrac{4 \tan \alpha}{1 - \tan^2 \alpha}}{1 - \dfrac{4 \tan^2 \alpha}{(1 - \tan^2 \alpha)^2}}$$

$$= \frac{\dfrac{4 \tan \alpha}{1 - \tan^2 \alpha}}{\dfrac{1 - 2 \tan^2 \alpha + \tan^4 \alpha - 4 \tan^2 \alpha}{(1 - \tan^2 \alpha)^2}}$$

$$= \frac{4 \tan \alpha \ (1 - \tan^2 \alpha)}{1 - 6 \tan^2 \alpha + \tan^4 \alpha} \qquad \text{Q.E.D}$$

Solution-22: Working on LHS of equation, we have

$$\cos 4\beta = \cos (2\beta + 2\beta) = \cos 2\beta \cos 2\beta - \sin 2\beta \sin 2\beta$$

Applying the double angle identities, we have

$$= (\cos^2 \beta - \sin^2 \beta)(\cos^2 \beta - \sin^2 \beta) - (2 \sin \beta \cos \beta)^2$$

$$= (\cos^2 \beta - \sin^2 \beta)(\cos^2 \beta - \sin^2 \beta) - 4 \sin^2 \beta \cos^2 \beta$$

$$= \cos^4 \beta + \sin^4 \beta - 2 \cos^2 \beta \sin^2 \beta - 4 \sin^2 \beta \cos^2 \beta$$

$$= (\cos^2 \beta + \sin^2 \beta)^2 - 4 \cos^2 \beta \sin^2 \beta - 4 \sin^2 \beta \cos^2 \beta$$

$$= 1 - 8 \sin^2 \beta \cos^2 \beta \qquad \text{Q.E.D}$$

Alternatively, we could solve as below:

$$\cos 4\beta = \cos(2 \cdot 2\beta) = 1 - 2 \sin^2 2\beta = 1 - 2(2 \sin \beta \cos \beta)^2$$

$$= 1 - 2(4 \sin^2 \beta \cos^2 \beta) = 1 - 8 \sin^2 \beta \cos^2 \beta \qquad \text{Q.E.D}$$

Solution-23: We expand the expression using sum identity as

$$\cot 3\beta = \cot(2\beta + \beta)$$

$$= \frac{\cot 2\beta \cot \beta - 1}{\cot 2\beta + \cot \beta}$$

Rearranging the complete equation as,

$$\cot 3\beta (\cot 2\beta + \cot \beta) = \cot 2\beta \cot \beta - 1$$

$$\cot 3\beta \cot 2\beta + \cot 3\beta \cot \beta = \cot 2\beta \cot \beta - 1$$

Arranging the equation to bring all cot expression to LHS and 1 at RHS,

$$\cot 2\beta \cot \beta - \cot 3\beta \cot 2\beta - \cot 3\beta \cot \beta = 1 \qquad \text{QED}$$

5 Chapter 5: Solving Trigonometric Equations

5.1 Trigonometric Equation

An equation involving trigonometric functions is trigonometric equation. Solving a trigonometric equation means to find a set of all values of an unknown angle, which satisfies the given equation.

Let's consider a trigonometric equation:

$$\sin \theta = \frac{1}{2}.$$

A solution of given trigonometric equation is the value of the unknown angle θ that satisfies the equation. This equation is satisfied by $\theta = \dfrac{-7\pi}{6}, \dfrac{\pi}{6}, \dfrac{5\pi}{6}, \dfrac{13\pi}{6}, \dfrac{17\pi}{6}$ etc. so these are solutions of the equation besides infinitely many other solutions. But how do we represent so many solutions and under which category?

There are two categories to represent the solutions: principal solution and general solution.

5.2 Principal Solution

For given trigonometric equation:

$$\sin \theta = \frac{1}{2}$$

Specific solutions of trigonometric equation for θ in the interval [0, 2π) are called principal solution. The interval [0, 2π) of θ specifies $0 \le \theta < 2\pi$, where angle 0 is included but 2π is excluded.

We observe that principal solution of the equation $\sin \theta = \dfrac{1}{2}$ are $\theta = \dfrac{\pi}{6}$ and $\theta = \dfrac{5\pi}{6}$, because these solutions lie between 0 to 2π.

Note that the first principal solution is always the smallest positive angle.

Example-1: Find the principal solution of equation
$$\sin \beta = \frac{\sqrt{3}}{2}$$

Solution: As $\sin \dfrac{\pi}{3} = \dfrac{\sqrt{3}}{2}$, so $\beta = \dfrac{\pi}{3}$ is first principal solution.

From supplementary identity, we find other principal solution in the interval $[0, 2\pi)$

$$\sin \frac{\pi}{3} = \sin \left(\pi - \frac{\pi}{3}\right) = \sin \frac{2\pi}{3}$$

Therefore, principal solutions are $\beta = \dfrac{\pi}{3}$ and $\dfrac{2\pi}{3}$

Example-2: Find the principal solution of equation
$$\tan \beta = -\frac{1}{\sqrt{3}}$$

Solution: As $\tan \dfrac{\pi}{6} = \dfrac{1}{\sqrt{3}}$, so from supplementary identity we have

$$-\tan \frac{\pi}{6} = \tan \left(\pi - \frac{\pi}{6}\right) = \tan \frac{5\pi}{6}$$

Therefore, first principal solution is $\beta = \dfrac{5\pi}{6}$

To find another solution, we have

$$-\tan\frac{\pi}{6} = \tan\left(2\pi - \frac{\pi}{6}\right) = \tan\frac{11\pi}{6}$$

So other principal solution is $\dfrac{11\pi}{6}$

Therefore, principal solutions are $\beta = \dfrac{5\pi}{6}$ and $\dfrac{11\pi}{6}$

Example-3: Find the principal solution of equation
$$\sin\beta = -\frac{1}{\sqrt{2}}$$

Solution: As $\sin\dfrac{\pi}{4} = \dfrac{1}{\sqrt{2}}$, so from $(\pi+\theta)$ identity we have

$$-\sin\frac{\pi}{4} = \sin\left(\pi + \frac{\pi}{4}\right) = \sin\frac{5\pi}{4}$$

Therefore, first principal solution is $\beta = \dfrac{5\pi}{4}$

To find another solution, we use

$$-\sin\frac{\pi}{4} = \sin\left(2\pi - \frac{\pi}{4}\right) = \sin\frac{7\pi}{4}$$

So other principal solution is $\dfrac{7\pi}{4}$

Therefore, principal solutions are $\beta = \dfrac{5\pi}{4}$ and $\dfrac{7\pi}{4}$

Example-4: Find the principal solution of equation
$$\cos\beta = -\frac{1}{\sqrt{2}}$$

Solution: As $\cos\dfrac{\pi}{4} = \dfrac{1}{\sqrt{2}}$, so from $(\pi-\theta)$ identity we have

$$-\cos\frac{\pi}{4} = \cos\left(\pi - \frac{\pi}{4}\right) = \cos\frac{3\pi}{4}$$

Therefore, first principal solution is $\beta = \dfrac{3\pi}{4}$

To find another solution, we use

$$-\cos\frac{\pi}{4} = \cos\left(\pi + \frac{\pi}{4}\right) = \cos\frac{5\pi}{4}$$

So other principal solution is $\beta = \dfrac{5\pi}{4}$

Therefore, principal solutions are $\beta = \dfrac{3\pi}{4}$ and $\dfrac{5\pi}{4}$

5.3   General Solution

A solution generalized by means of periodicity is known as the general solution. The solutions involving integer n, which gives all solutions of a trigonometric equation, are generalized in form of a general solution.

Let's consider an equation,

$$2\cos\theta + 1 = 0$$

The equation is simplified as $\cos\theta = -\dfrac{1}{2}$, which is satisfied by $\theta = \dfrac{-4\pi}{3}, \dfrac{-2\pi}{3}, \dfrac{2\pi}{3}, \dfrac{4\pi}{3}$ etc. Since the trigonometric functions are periodic, therefore, if a trigonometric equation has a solution, it will have infinite number of solutions.

As discussed earlier, sin and cos repeats itself after interval of 2π

but tan repeats after π. We will use periodicity in solving trigonometric equation.

To find general solution, we find the first principal solution (smallest positive angle) and thereafter apply periodicity.

Therefore, $\theta = \dfrac{2\pi}{3}$, $2\pi \pm \dfrac{2\pi}{3}$, $4\pi \pm \dfrac{2\pi}{3}$, ... are solutions of $2\cos\theta + 1 = 0$. These solutions can be put together in compact and generic form as $2n\pi \pm \dfrac{2\pi}{3}$, where n is an integer.

This solution $\theta = 2n\pi \pm \dfrac{2\pi}{3}$ for given equation is known as the general solution.

Theorems for General Solutions

Theorem 5.1: For any angles α and β,
$\sin\alpha = \sin\beta$ implies $\alpha = n\pi + (-1)^n \beta$
    where $n \in Z$ *(a set of all integers); and*
    β *is the first principal solution (smallest positive angle).*

Example-1: Use the theorem 5.1 to find the general solution of equation

$$\sin\beta = -\frac{\sqrt{3}}{2}$$

Solution: $-\dfrac{\sqrt{3}}{2} = -\sin\dfrac{\pi}{3} = \sin(\pi + \dfrac{\pi}{3}) = \sin\dfrac{4\pi}{3}$

As $\dfrac{4\pi}{3}$ is the first principal solution (smallest positive angle), So general solution is,

$$\beta = n\pi + (-1)^n \frac{4\pi}{3}, \text{ where } n \in Z$$

Theorem 5.2: For any angles α and β,
$\cos \alpha = \cos \beta$ implies $\alpha = 2n\pi \pm \beta$
> *where* $n \in Z$ *(a set of all integers); and*
> β *is the first principal solution (smallest positive angle).*

Exmaple-2: Use the theorem 5.2 to find the general solution of equation
$$\cos \beta = \frac{1}{2}$$

Solution: We have $\cos \frac{\pi}{3} = \frac{1}{2}$

As $\beta = \frac{\pi}{3}$ is the first principal solution (smallest positive angle), So general solution is,

$$\beta = 2n\pi \pm \frac{\pi}{3}, \text{ where } n \in Z$$

Theorem 5.3: If angles α and β are not odd multiple of $\frac{\pi}{2}$, then
$\tan \alpha = \tan \beta$ implies $\alpha = n\pi + \beta$
> *where* $n \in Z$ *(a set of all integers); and*
> β *is the first principal solution (smallest positive angle).*

Example-3: Use the theorem 5.3 to solve the equation,
$$\tan 2\alpha = -\cot \left(\alpha + \frac{\pi}{3}\right)$$

Solution: $\tan 2\alpha = -\cot \left(\alpha + \frac{\pi}{3}\right) = \tan \left(\frac{\pi}{2} + \alpha + \frac{\pi}{3}\right) = \tan \left(\alpha + \frac{5\pi}{6}\right)$

As per theorem of general solution of tan,

$$2\alpha = n\pi + \alpha + \frac{5\pi}{6} \text{ , where } n \in Z$$

Therefore,

$$\alpha = n\pi + \frac{5\pi}{6} \text{ , where } n \in Z$$

Theorem 5.4: For any angles α and β,

$\sin^2 \alpha = \sin^2 \beta$ implies $\alpha = n\pi \pm \beta$

> *where* $n \in Z$ *(a set of all integers); and*
> β *is the first principal solution (smallest positive angle).*

Theorem 5.5: For any angles α and β,

$\cos^2 \alpha = \cos^2 \beta$ implies $\alpha = n\pi \pm \beta$

> *where* $n \in Z$ *(a set of all integers); and*
> β *is the first principal solution (smallest positive angle).*

Theorem 5.6: For any angles α and β,

$\tan^2 \alpha = \tan^2 \beta$ implies $\alpha = n\pi \pm \beta$

> *where* $n \in Z$ *(a set of all integers); and*
> β *is the first principal solution (smallest positive angle).*

Example-4: Solve the equation $\cos^2 \alpha = \frac{1}{2}$

Solution: $\cos^2 \alpha = \frac{1}{2}$ $=>$ $\cos^2 \alpha = \left(\frac{1}{\sqrt{2}} \right)^2$

$=>$ $\cos^2 \alpha = \cos^2 \frac{\pi}{4}$

Applying the theorem 5.5, we have the general solution

$\alpha = n\pi \pm \dfrac{\pi}{4}$, where $n \in Z$

General Solutions for Specific Values

The general solutions of equations given below can be verified using theorem 5.1, theorem 5.2 or theorem 5.3.

Case-1: $\sin \alpha = 0$ $\Rightarrow \alpha = n\pi$

Case-2: $\sin \alpha = 1$ $\Rightarrow \alpha = (4n+1)\dfrac{\pi}{2}$

Case-3: $\sin \alpha = -1$ $\Rightarrow \alpha = (4n-1)\dfrac{\pi}{2}$

Case-4: $\cos \alpha = 0$ $\Rightarrow \alpha = (2n+1)\dfrac{\pi}{2}$

Case-5: $\cos \alpha = 1$ $\Rightarrow \alpha = 2n\pi$

Case-6: $\cos \alpha = -1$ $\Rightarrow \alpha = (2n+1)\pi$

Case-7: $\tan \alpha = 0$ $\Rightarrow \alpha = n\pi$

where **n** $\in$ **Z** *(set of all integers)*

Example-5: Solve the equation

$$2 \cos^2 \alpha + 3 \sin \alpha = 0$$

Solution: The equation can be simplified as,

$$2 (1 - \sin^2 \alpha) + 3 \sin \alpha = 0$$

Simplifying further,
$$2 - 2 \sin^2 \alpha + 3 \sin \alpha = 0$$

Multiplying each side by -1, we have
$$2 \sin^2 \alpha - 3 \sin \alpha - 2 = 0$$

Factorizing the given equation, we get
$$(2 \sin \alpha + 1)(\sin \alpha - 2) = 0$$

The equation above implies that either the term $(2 \sin \alpha + 1) = 0$, or the term $(\sin \alpha - 2) = 0$.

The term $(\sin \alpha - 2)$ of equation implies that $\sin \alpha = 2$, which is invalid, because value of sin function can never be greater than 1. Therefore, there is no solution for this term.

The term $(2 \sin \alpha + 1)$ of equation implies that $\sin \alpha = \dfrac{-1}{2}$, whose first principal solution is $\alpha = \dfrac{7\pi}{6}$

Therefore, the general solution is
$$\alpha = n\pi + (-1)^n \, \frac{7\pi}{6}, \text{ where } n \in Z$$

Worksheet-14

Exercise-1: Write the general solution for $\sin \alpha = \sin \beta$, where β is the first principal solution.

Exercise-2: Write the general solution for $\cos \alpha = \cos \beta$, where β is the first principal solution.

Exercise-3: Write the general solution for $\tan \alpha = \tan \beta$, where β is the first principal solution.

Exercise-4: Write the general solution for $\sin^2 \alpha = \sin^2 \beta$, where β is the first principal solution.

Exercise-5: Find principal and general solutions for $\cos \beta = \dfrac{1}{2}$

Exercise-6: Find principal and general solution for $\sin \beta = \dfrac{1}{2}$

Exercise-7: Find principal and general solutions for $tan\ \beta = -1$

Exercise-8: Find principal and general solutions for $cot\ \beta = -\sqrt{3}$

Exercise-9: Find principal and general solutions for $cosec\ \beta = -2$

Exercise-10: Verify following

Case-1: $\sin \alpha = 0 \quad => \alpha = n\pi$

Case-2: $\sin \alpha = 1 \quad \Rightarrow \alpha = (4n+1)\dfrac{\pi}{2}$

Case-3: $\sin \alpha = -1 \quad \Rightarrow \alpha = (4n-1)\dfrac{\pi}{2}$

Case-4: $\cos \alpha = 0 \quad \Rightarrow \alpha = (2n+1)\dfrac{\pi}{2}$

Case-5: $\cos \alpha = 1 \quad \Rightarrow \alpha = 2n\pi$

Case-6: $\cos \alpha = -1 \quad \Rightarrow \alpha = (2n +1)\pi$

Case-7: $\tan \alpha = 0 \quad \Rightarrow \alpha = n\pi$

Exercise-11: Solve

$$\sin 8\beta + \sin 2\beta - \sin 5\beta = 0$$

Exercise-12: Solve

$$2 \cos^2 \beta + 7 \sin \beta + 2 = 0$$

Solutions-Worksheet-14

Solution-1: $\alpha = n\pi + (-1)^n \beta$

Solution-2: $\alpha = 2n\pi \pm \beta$

Solution-3: $\alpha = n\pi + \beta$

Solution-4: $\alpha = n\pi \pm \beta$

Solution-5: As $\dfrac{1}{2} = \cos\dfrac{\pi}{3}$ and $\cos\left(2\pi - \dfrac{\pi}{3}\right) = \cos\left(\dfrac{5\pi}{3}\right)$

Therefore, principal solutions are $\beta = \dfrac{\pi}{3}$ and $\dfrac{5\pi}{3}$

Applying the theorem 5.2 for cos α = cos β, the general solution is $\alpha = 2n\pi \pm \beta$;

Hence, the general solution of $\cos\beta = \dfrac{1}{2}$ is

$$\beta = 2n\pi \pm \dfrac{\pi}{3}$$

Solution-6: As $\sin\dfrac{\pi}{6} = \dfrac{1}{2}$, and $\sin\left(\pi - \dfrac{\pi}{6}\right) = \sin\dfrac{5\pi}{6}$

Therefore, principal solutions are $\beta = \dfrac{\pi}{6}$ and $\dfrac{5\pi}{6}$

Applying theorem 5.1 for sin α = sin β, the general solution is $\alpha = n\pi + (-1)^n \beta$;

Hence, the general solution of $\sin\beta = \dfrac{1}{2}$ is

$$\beta = n\pi + (-1)^n \dfrac{\pi}{6}$$

Solution-7: As $\tan\dfrac{\pi}{4} = 1$, so its negative value is

$-\tan\dfrac{\pi}{4} = \tan\left(\pi - \dfrac{\pi}{4}\right) = \tan\dfrac{3\pi}{4}$

Therefore, first principal solution is $\beta = \dfrac{3\pi}{4}$

To find another solution, we have

$$-\tan\frac{\pi}{4} = \tan\left(2\pi - \frac{\pi}{4}\right) = \tan\frac{7\pi}{4}$$

So, other principal solution is $\beta = \dfrac{7\pi}{4}$

Applying the theorem 5.3 for *tan α = tan β, the general solution is* $\alpha = n\pi + \beta;$ hence the general solution of *tan β = −1* is

$$\beta = n\pi + \frac{3\pi}{4}$$

Solution-8: We transform the equation in terms of tan, so

$$\tan\beta = -\frac{1}{\sqrt{3}}$$

As $\tan\dfrac{\pi}{6} = \dfrac{1}{\sqrt{3}}$, so its negative value is

$$-\tan\frac{\pi}{6} = \tan\left(\pi - \frac{\pi}{6}\right) = \tan\frac{5\pi}{6}$$

Therefore, first principal solution is $\beta = \dfrac{5\pi}{6}$

To find another solution, we have

$$-\tan\frac{\pi}{6} = \tan\left(2\pi - \frac{\pi}{6}\right) = \tan\frac{11\pi}{6}$$

So, other principal solution is $\beta = \dfrac{11\pi}{6}$

Applying the theorem 5.3 for *tan α = tan β, the general solution is* $\alpha = n\pi + \beta;$ hence the general solution of *tan β = −1* is

$$\beta = n\pi + \frac{5\pi}{6}$$

Solution-9: As $\sin \beta = -\dfrac{1}{2}$,

As $\sin \dfrac{\pi}{6} = \dfrac{1}{2}$, so its negative value is

$$-\sin \dfrac{\pi}{6} = \sin \left(\pi + \dfrac{\pi}{6}\right) = \sin \dfrac{7\pi}{6}$$

Therefore, first principal solution is $\beta = \dfrac{7\pi}{6}$

To find another solution, we have

$$-\sin \dfrac{\pi}{6} = \sin \left(2\pi - \dfrac{\pi}{6}\right) = \sin \dfrac{11\pi}{6}$$

So, other principal solution is $\beta = \dfrac{11\pi}{6}$

Applying theorem 5.1 for $\sin \alpha = \sin \beta$, *the general solution is* $\alpha = n\pi + (-1)^n \beta$; hence the general solution of $\sin \beta = \dfrac{-1}{2}$ is

$$\beta = n\pi + (-1)^n \dfrac{7\pi}{6}$$

Solution-10-Case-1:
As per theorem 5.1: $\sin \alpha = \sin \beta$ implies $\alpha = n\pi + (-1)^n \beta$
We have $\beta = 0$, so
$\alpha = n\pi + (-1)^n \, 0$; this implies $\alpha = n\pi$

Solution-10-Case-2:
As per theorem 5.1: $\sin \alpha = \sin \beta$ implies $\alpha = n\pi + (-1)^n \beta$
We have $\beta = \dfrac{\pi}{2}$, so
$\alpha = n\pi + (-1)^n \dfrac{\pi}{2}$; this implies $\alpha = (4n+1)\dfrac{\pi}{2}$

Solution-10-Case-3:
As per theorem 5.1: $\sin \alpha = \sin \beta$ implies $\alpha = n\pi + (-1)^n \beta$

We have $\beta = (\pi + \frac{\pi}{2})$, so

$\alpha = n\pi + (-1)^n \frac{3\pi}{2}$; this implies $\alpha = (4n-1)\frac{\pi}{2}$

Solution-10-Case-4:

As per theorem 5.2: $\cos \alpha = \cos \beta$ implies $\alpha = 2n\pi \pm \beta$

We have $\beta = \frac{\pi}{2}$, so

$\alpha = 2n\pi \pm \frac{\pi}{2}$; this implies $\alpha = (2n+1)\frac{\pi}{2}$

Solution-10-Case-5:

As per theorem 5.2: $\cos \alpha = \cos \beta$ implies $\alpha = 2n\pi \pm \beta$

We have $\beta = 0$, so

$\alpha = 2n\pi \pm 0$; this implies $\alpha = 2n\pi$

Solution-10-Case-6:

As per theorem 5.2: $\cos \alpha = \cos \beta$ implies $\alpha = 2n\pi \pm \beta$

We have $\beta = \pi$, so

$\alpha = 2n\pi \pm \pi$; this implies $\alpha = (2n+1)\pi$

Solution-10-Case-7:

As per theorem 5.3: $\tan \alpha = \tan \beta$ implies $\alpha = n\pi + \beta$

We have $\beta = 0$, so

$\alpha = n\pi + 0$; this implies $\alpha = n\pi$

Solution-11:

Applying sum as product identity,　we get

$\sin 8\beta + \sin 2\beta = 2 \sin 5\beta \cos 3\beta$

So simplifying the LHS of equation, we have

$2 \sin 5\beta \cos 3\beta - \sin 5\beta = 0$

$\sin 5\beta (2 \cos 3\beta - 1) = 0$

Therefore,

$$\sin 5\beta = 0 \text{ or } \cos 3\beta = \frac{1}{2}$$

Solving for $\sin 5\beta=0$, the general solution is:

$$5\beta = n\pi, \text{ so } \beta = \frac{n\pi}{5}, \text{ where } n \in Z$$

Solving for $\cos 3\beta = \frac{1}{2}$, we have

$$\cos 3\beta = \cos \frac{\pi}{3}, \text{ so the general solution is:}$$

$$3\beta = 2n\pi \pm \frac{\pi}{3}, \text{ so } \beta = \frac{2n\pi}{3} \pm \frac{\pi}{9} \text{ where } n \in Z$$

Solution-12: LHS of equation is simplified as,

$$2\,(1 - \sin^2 \beta) + 7 \sin \beta + 2 = 0$$

Multiplying by –1 to both sides of equation,

$$2 \sin^2 \beta - 7 \sin \beta - 4 = 0$$
$$(2 \sin \beta + 1)\,(\sin \beta - 4) = 0$$

Hence, $\sin \beta = -\frac{1}{2}$ or $\sin \beta = 4$

As the value of sin cannot be more than 1, so *sin β=4* does not have any solution.

Solving the term $\sin \beta = -\frac{1}{2}$, we have

$$\sin \beta = -\frac{1}{2} = \sin \left(\pi + \frac{\pi}{6}\right) = \sin \frac{7\pi}{6}$$

Therefore, $\beta = n\pi + (-1)^n \frac{7\pi}{6}$, where $n \in Z$

Chapter 6: Inverse Trigonometric Functions

6.1 Concept of Inverse Trigonometric Function

The inverse trigonometric functions are the inverse of the trigonometric functions with restricted range in the principal branch.

Notation of Inverse Trigonometric Function

The inverse trigonometric functions are denoted by notations $\sin^{-1}(\beta)$, $\cos^{-1}(\beta)$, $\tan^{-1}(\beta)$ etc. $\sin^{-1}(\beta)$ is read as "inverse Sine of β". In this notation, the -1 over Sine looks like an exponent but it isn't, rather it actually denotes the inverse function. Therefore,

$$\sin^{-1}(\beta) \neq \frac{1}{\sin(\beta)}$$

To use -1 as exponent, we will denote it as $(\sin \beta)^{-1}$, so

$$(\sin \beta)^{-1} = \frac{1}{\sin(\beta)}$$

Remark: Another convention is to name inverse trigonometric functions using an arc- prefix, like arcsin(x), arccos(x), arctan(x), etc., but this convention is not used in this book.

Concept of Inverse Trigonometric Function

The inverse trigonometric function is used to determine an angle.

Expression $\sin^{-1} \frac{1}{\sqrt{2}}$ denotes the angle whose Sine is equal to $\frac{1}{\sqrt{2}}$

For an equation, $\sin^{-1} x = \beta$; Sine of angle β is the value x.

Similarly, for an equation: $\cos^{-1} y = \alpha$, Cosine of angle α is the value y.

Remark: Note that the non-inverse function $y = sin\ x$ is also expressed as forward trigonometric function. We will often use this term in the chapter.

Example-1: Evaluate $\sin^{-1} \dfrac{1}{\sqrt{2}}$

Solution: Let's assume $\sin^{-1} \dfrac{1}{\sqrt{2}} = \beta$

In an equivalent equation $\sin \beta = \dfrac{1}{\sqrt{2}}$, angle β is $\dfrac{\pi}{4}$ whose Sine produces the value $\dfrac{1}{\sqrt{2}}$, therefore

$$\sin^{-1} \dfrac{1}{\sqrt{2}} = \dfrac{\pi}{4}$$

Example-2: Write the equivalent inverse function for $cos\ x = A$

Solution: The equivalent inverse function is $\cos^{-1} A = x$

Example-3: Write the equivalent non-inverse (forward) function for $sin^{-1} 1 = \beta$

Solution: The equivalent forward function is $sin\ \beta = 1$

Example-4: Write the equivalent non-inverse (forward) function for $tan^{-1} 1 = \frac{\pi}{4}$

Solution: The equivalent forward function is $tan \frac{\pi}{4} = 1$

6.2 Relationships of Inverse Trigonometric Functions

The Inverse Relations

Let's assume a forward function as $f(y) = sin\ y$ and another inverse function as $g(y) = sin^{-1} y$

As per definition of inverse function,

$f(g(y)) = y$, and $g(f(y)) = y.$

Hence, putting in the values of $f(g(y))$ and $g(f(y))$, we have

$sin\ (sin^{-1} y) = y$, and $sin^{-1}(sin\ y) = y$

Example-1: Evaluate $sin^{-1}(sin \frac{\pi}{4})$

Solution: As per inverse function relation, $sin^{-1}(sin\ y) = y.$

Therefore, $sin^{-1}(sin \frac{\pi}{4}) = \frac{\pi}{4}$

Alternatively, we can evaluate using substituting the appropriate value, without using inverse relation.

$$sin^{-1}(sin\frac{\pi}{4}) = sin^{-1}(\frac{1}{\sqrt{2}}) = \frac{\pi}{4}$$

Example-2: Evaluate $cos\ (cos^{-1}(\frac{\sqrt{3}}{2}))$

Solution: As per inverse function relation, $cos\ (cos^{-1}y) = y$.

Therefore, $cos\ (cos^{-1}(\frac{\sqrt{3}}{2})) = \frac{\sqrt{3}}{2}$

Example-3: Solve for x

$$cos^{-1}(x + 1) = \frac{\pi}{4}$$

Solution: As per definition of inverse function,

$$x + 1 = cos\frac{\pi}{4}$$

$$x = \frac{1}{\sqrt{2}} - 1 = \frac{1-\sqrt{2}}{\sqrt{2}} = \frac{-0.414}{1.414} = -0.293$$

Relations between Inverse Functions

Inv-6.2.1:

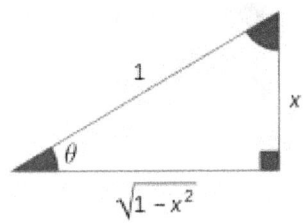

$$\sin^{-1}(x) = \cos^{-1}(\sqrt{1-x^2}) = \tan^{-1}\left(\frac{x}{\sqrt{1-x^2}}\right) = \cot^{-1}\left(\frac{\sqrt{1-x^2}}{x}\right)$$

$$= \sec^{-1}\left(\frac{1}{\sqrt{1-x^2}}\right) = \csc^{-1}\left(\frac{1}{x}\right)$$

Inv-6.2.2:

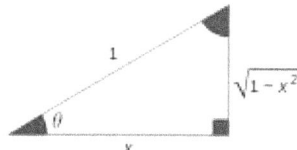

$$\cos^{-1}(x) = \sin^{-1}(\sqrt{1-x^2}) = \tan^{-1}\left(\frac{\sqrt{1-x^2}}{x}\right) = \cot^{-1}\left(\frac{x}{\sqrt{1-x^2}}\right)$$

$$= \sec^{-1}\left(\frac{1}{x}\right) = \csc^{-1}\left(\frac{1}{\sqrt{1-x^2}}\right)$$

Inv-6.2.3:

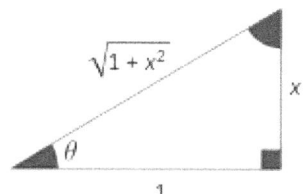

$$\tan^{-1}(x) = \sin^{-1}\left(\frac{x}{\sqrt{1+x^2}}\right) = \cos^{-1}\left(\frac{1}{\sqrt{1+x^2}}\right) = \cot^{-1}\left(\frac{1}{x}\right)$$

$$= \sec^{-1}(\sqrt{1+x^2}) = \csc^{-1}\left(\frac{\sqrt{1+x^2}}{x}\right)$$

6.3   Principal Branch, Range & Domain

Range and Principal Branch of Trigonometric Functions

Because of repetitious nature of trigonometric function, the essential information in any trigonometric function is contained in a very small and invertible part. These essential pieces are called the principal branches of function, as depicted in figure below for non-inverse trigonometric functions.

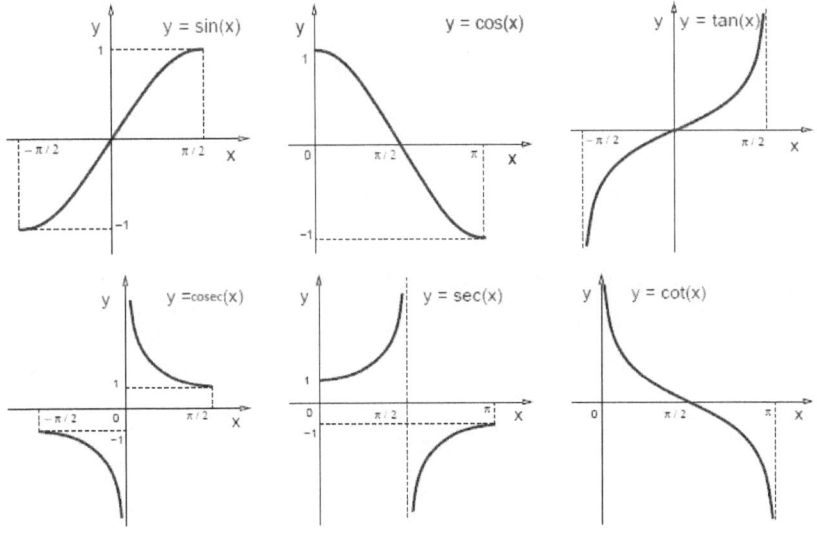

Conversely, the inverse function has the principal branch corresponding to non-inverse functions, as shown above.

For inverse trigonometric function, range is limited to and defined only in principal branch. Principal branch of inverse trigonometric functions are depicted in figure below.

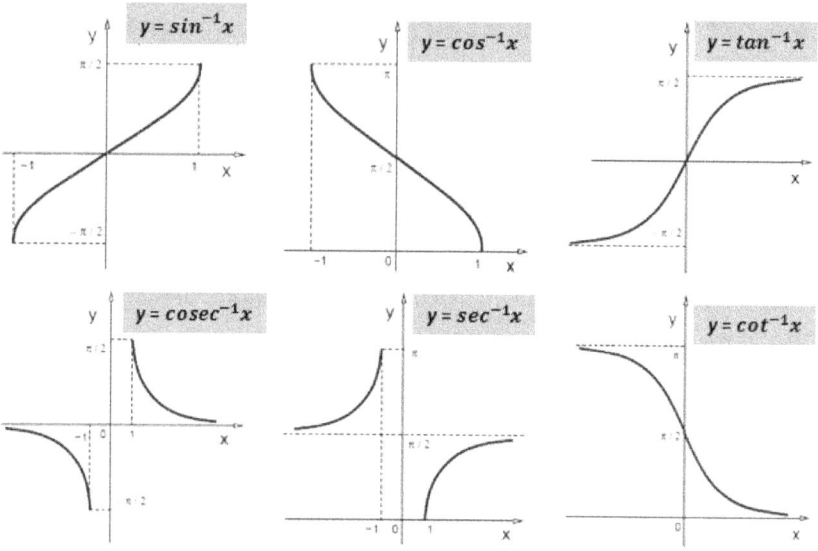

As per global convention, only the principal branches of the trigonometric functions are considered when finding inverses. It is noticeable in the figure that, in all cases, the principal branch is either strictly increasing or strictly decreasing, i.e. it's invertible. The range of outputs on the branch covers the entire range of outputs for the function itself.

For all of the inverse trigonometric functions, whole principle branch (principal range) can be described in a set of $[-\frac{\pi}{2}, \pi]$.

For all possible values of domain which an inverse trigonometric function might have, there is an angle which is of minimum absolute value, called principal value.

Principal Value

The angle with smallest absolute value in the principal branch is called the principal value of inverse function.

Alternatively, the smallest absolute value of the inverse trigonometric function (an angle) which lies in the principal branch is its principal value.

Let's consider an equation, $sin^{-1} \frac{1}{\sqrt{2}} = x$

As $sin \frac{\pi}{4} = \frac{1}{\sqrt{2}}$

So, $x = \frac{\pi}{4}$ is the solution of the equation, but $\frac{\pi}{4}$ is not the only solution for x.

In second quadrant, Sine of angle $(\pi - \frac{\pi}{4})$ also has the value $\frac{1}{\sqrt{2}}$

$sin \frac{3\pi}{4} = \frac{1}{\sqrt{2}}$

Moreover, Sine is periodic with 2π. So, Sine of angle $(2\pi + \frac{\pi}{4})$ also equals $\frac{1}{\sqrt{2}}$

$sin (\frac{9\pi}{4}) = \frac{1}{\sqrt{2}}$

To solve this problem of having several solutions, we restrict the angle to principal value. The angle with smallest absolute value in the principal branch is called the principal value of inverse function.

So the solution for $sin^{-1} \frac{1}{\sqrt{2}}$ is $\frac{\pi}{4}$, as angle $\frac{\pi}{4}$ is the principal value (smallest absolute value in the principal branch).

<u>Example-1</u>: Evaluate $sin^{-1} (-\frac{1}{\sqrt{2}})$

Solution: In third and fourth quadrant, value of Sine is negative. Angles $-\frac{3\pi}{4}$ in third quadrant and $-\frac{\pi}{4}$ in fourth quadrant are the angles whose Sines produce the same value $-\frac{1}{\sqrt{2}}$.

But the angle of smallest absolute value is $-\frac{\pi}{4}$, which lies in the principal branch of inverse Sine $[-\frac{\pi}{2}, \frac{\pi}{2}]$.

Therefore, $\frac{-\pi}{4}$ is the principal value.

Example-2: Evaluate principal value of $cos^{-1}(-\frac{\sqrt{3}}{2})$

Solution: The corresponding acute angle for value $\frac{\sqrt{3}}{2}$ of Cosine is $\frac{\pi}{6}$.

As the value is negative, Cosine must be in second or third quadrant. Principal branch (range) of Cosine is $[0, \pi]$. So we must select the angle in second quadrant. Second quadrant angle whose corresponding acute angle is $\frac{\pi}{6}$ must be $(\pi - \frac{\pi}{6}) = \frac{5\pi}{6}$

Alternatively second quadrant angle can also be found using supplementary identity of Cosine, which is *cos (180°–θ) = −cos θ*

So, $cos\ (\pi - \frac{\pi}{6}) = -\ cos\ \frac{\pi}{6}$. Therefore, $cos\ (\frac{5\pi}{6}) = -\frac{\sqrt{3}}{2}$

Hence, $\frac{5\pi}{6}$ is the principal value (smallest absolute value in the principal branch).

Domain and Range

For normal trigonometric function, like $sin\ x = y$; a set of all possible values of y is the range and a set of all possible values of x is the domain.

For inverse trigonometric equation, like $sin^{-1} y = x$, a set of all possible values of x is the range and a set of all possible values of y is the domain. If domain (y value) of an inverse function is positive, then the value of the inverse function is always a first quadrant angle. If domain (y value) is negative, the value of the inverse will fall in the quadrant in which the non-inverse (forward) function is negative.

For any inverse trigonometric equation, like $sin^{-1} y = x$, the range of $sin^{-1} y$ is equivalent to the domain of $sin\ x$. Similarly, the range of any inverse function is equivalent to the domain of corresponding forward (non-inverse) function.

Domain and Principal Branch of Inverse Sine

If Sine of an angle is positive, we choose the first quadrant angle; while for an angle whose Sine is negative, we choose the fourth quadrant angle.

From the discussion above, we observe that range x of equation $sin^{-1} y = x$ will be the angles that fall in first or fourth quadrant. Therefore,

If $sin^{-1} y = x$, then $-\dfrac{\pi}{2} \le x \le \dfrac{\pi}{2}$

For function $sin^{-1} y = x$, the principal branch (range) is $[-\frac{\pi}{2},$ $\frac{\pi}{2}]$ and domain is the interval $[-1, 1]$.

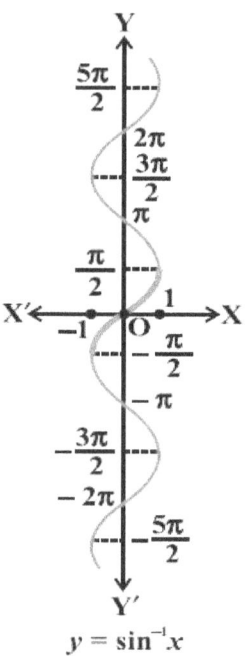

$$y = sin^{-1}x$$

Here, the range (x) or principal branch is $[-\frac{\pi}{2}, \frac{\pi}{2}]$ and domain (all possible y values) is the interval $[-1, 1]$.

Domain and range of Sine function:

For non-inverse function $y = \sin x$, where angle is not limited to principal branch, domain (all x values) is the set of all real numbers and range (all y values) is the interval $[-1, 1]$

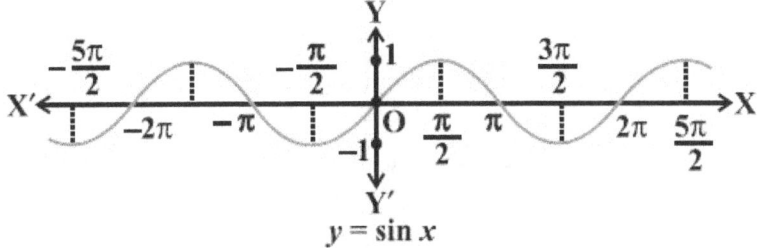

$$y = \sin x$$

Domain and Principal Branch of Inverse Tangent

If Tangent of an angle is positive, we choose the first quadrant angle; while for an angle whose Tangent is negative, we select the fourth quadrant angle.

Similar to inverse function of Sine, the inverse function of Tangent also has its smallest absolute value in first and fourth quadrants. For Tangent function, the quadrantal angles $\frac{\pi}{2}$ and $-\frac{\pi}{2}$ are not included. Therefore,

$$\text{If } tan^{-1} y = x, \text{ then } -\frac{\pi}{2} < x < \frac{\pi}{2}$$

For function $tan^{-1} y = x$, the principal-branch or range (values of x) is $(-\frac{\pi}{2}, \frac{\pi}{2})$; and domain (values of y) is a set of all real numbers $(-\infty, \infty)$.

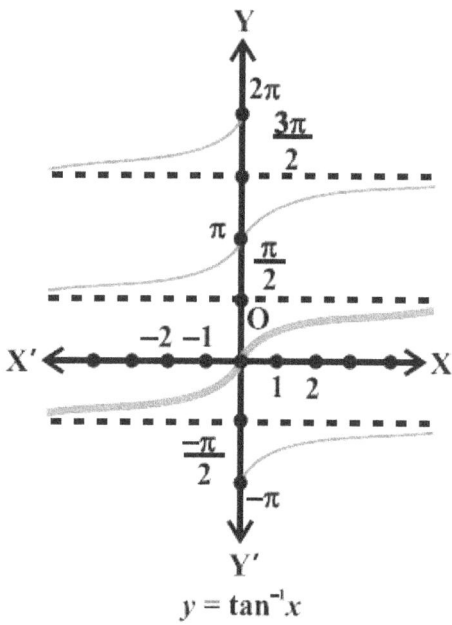

$$y = \tan^{-1}x$$

Remarks: Notice the difference between principal value domains of sin and tan. Principal value domain of y = sin x is closed interval $[-\frac{\pi}{2}, \frac{\pi}{2}]$, where $-\frac{\pi}{2}$ and $\frac{\pi}{2}$ are included. Principal value domain of y = tan x is open interval $(-\frac{\pi}{2}, \frac{\pi}{2})$, where $-\frac{\pi}{2}$ and $\frac{\pi}{2}$ are not included.

Domain and range of Tangent function:

For non-inverse function $y = \tan x$, if angle is not limited to principal value, then the domain (all x values) is the set $\{ x : x \in \mathbf{R}$ and $x \neq (2n+1)\frac{\pi}{2}, n \in \mathbf{Z} \}$ and range (all y values) is the set of all real numbers. Note that $(2n+1)\frac{\pi}{2}$ is odd multiple of $\frac{\pi}{2}$, which produces undefined value, when used as domain of tan.

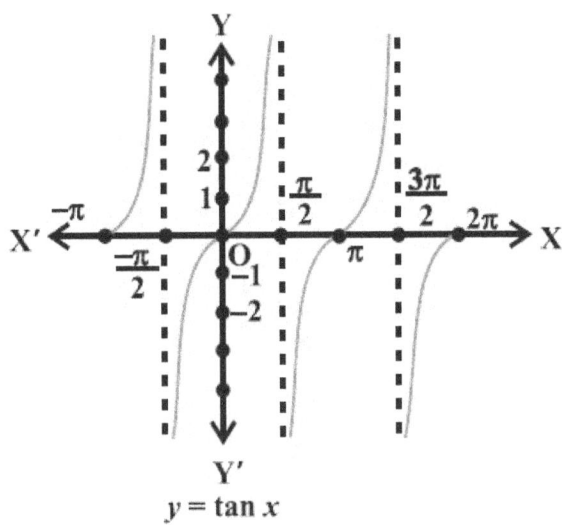

$y = \tan x$

Domain and Principal Branch of Inverse Cosine

For an angle whose Cosine is positive, we choose the first quadrant angle. Value of Cosine is negative in second and third quadrant, so we select second quadrant for negative value. The principal branch for Cosine is first and second quadrant.

The inverse Cosine of negative value is the angle which is supplementary of corresponding acute angle. Hence,

$$\cos^{-1}(-y) = \pi - \cos^{-1}(y)$$

This implies that the range of $\cos^{-1}(y)$ is from 0 to π. Therefore,

If $\cos^{-1} y = x$, then $0 \le x \le \pi$

For function $\cos^{-1} y = x$, the principal branch is [0, π] and domain is the interval [–1, 1]

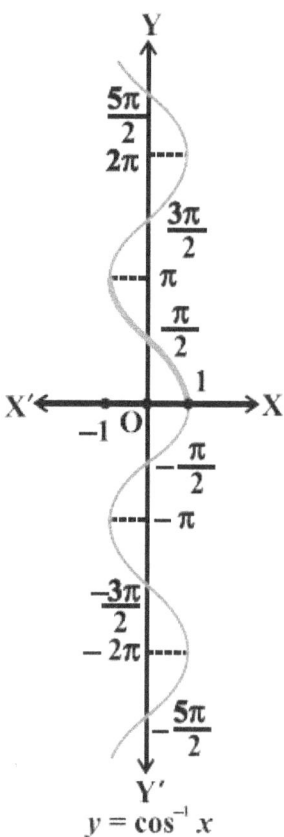

$$y = \cos^{-1} x$$

Domain and range of Cosine function:

For non-inverse function $y = \cos x$, if angle is not limited to principal value, domain (all possible x values) is the set of all real numbers $\{\ x : x \in R\ \}$ and range (all possible y values) is the interval $[-1, 1]$

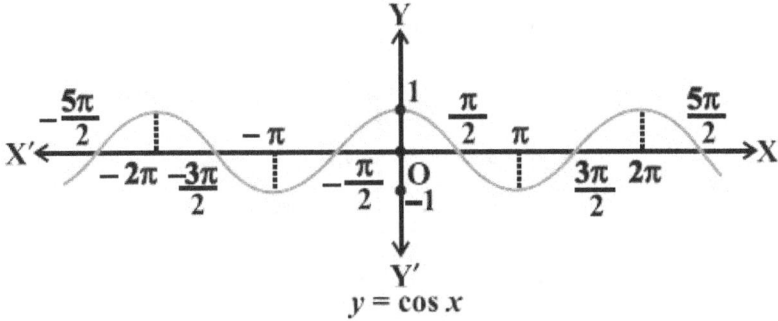

$$y = \cos x$$

Domain and Principal Branch of Inverse Secant

Trigonometric function of Secant is positive in first quadrant and negative in second quadrant.

The principal branch for inverse Secant is first and second quadrant.

If $sec^{-1} y = x$, then

$$0 \le x < \frac{\pi}{2} \text{ , for } y \ge 1$$

and $\frac{\pi}{2} < x \le \pi$, for $y \le -1$

For function $sec^{-1} y = x$, the principal branch is $[\ 0, \frac{\pi}{2})$ and ($\frac{\pi}{2}, \pi\ ]$; Domain is $y \ge 1$ and $y \le -1$.

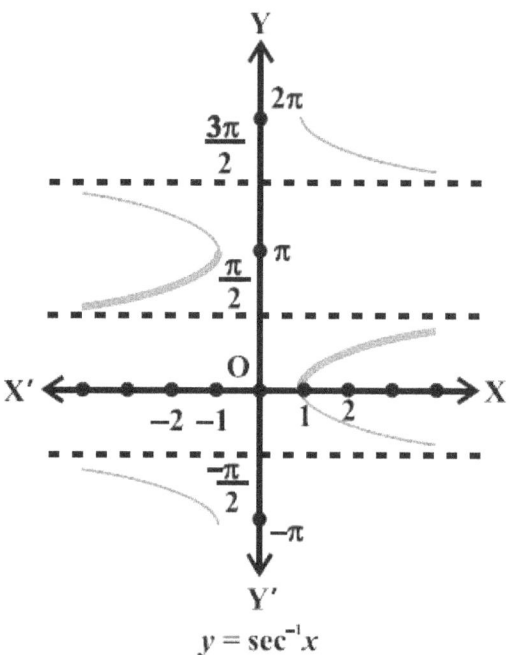

$$y = \sec^{-1}x$$

Domain/Range of Secant:

For non-inverse function y = sec x, if angle is not limited to principal value, the domain (all possible values of x) is the set { x : $x \in$ **R** and $x \neq (2n+1)\dfrac{\pi}{2}$, n $\in$ **Z** } and range (all possible values of y) is the set { y : y $\in$ **R,** y ≤ -1 or y ≥ 1 }

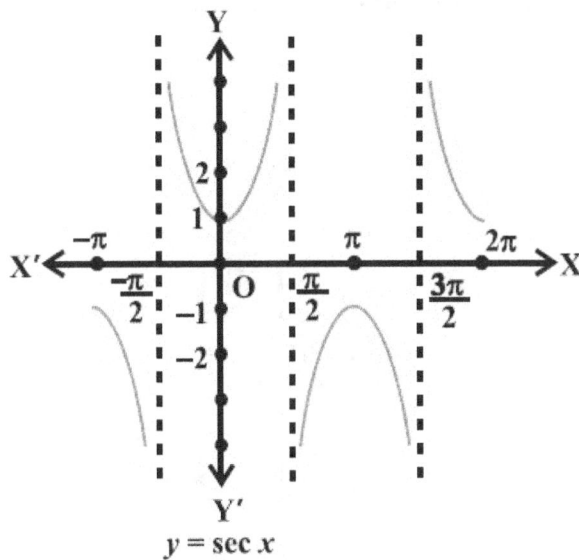

$$y = \sec x$$

Domain and Principal Branch of Inverse Cotangent

Trigonometric function of Cotangent is positive in first quadrant and negative in second quadrant.

The principal branch for inverse Cotangent is first and second quadrant.

If $cot^{-1} y = x$, then

$$0 < x \leq \frac{\pi}{2} \text{ , for } y \geq 0$$

$$\text{and } \frac{\pi}{2} < x < \pi \text{ , for } y < 0$$

For function $cot^{-1} y = x$, the principal branch is (0, π);

Domain is a set of all real numbers $(-\infty, \infty)$.

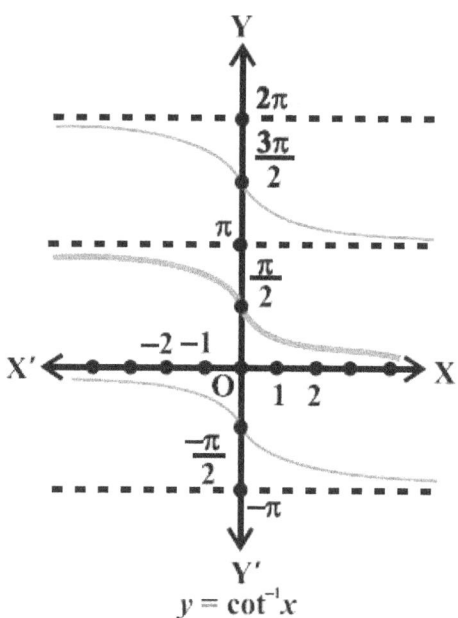

$$y = \cot^{-1}x$$

Remarks: For inverse function of Cotangent, some books and software-apps consider the principal branch as $(-\frac{\pi}{2}, 0)$ and $(0, \frac{\pi}{2}]$; but we will consider the principal branch as $(0, \pi)$ because cot inverse is continuous over the entire range $(0, \pi)$.

Domain/Range of Cotangent:

For non-inverse function $y = \cot x$, if angle is not limited to principal value, the domain (all possible values of x) is the set $\{ x : x \in R \text{ and } x \neq n\pi, n \in Z \}$ and range (all possible values of y) is the set of all real numbers.

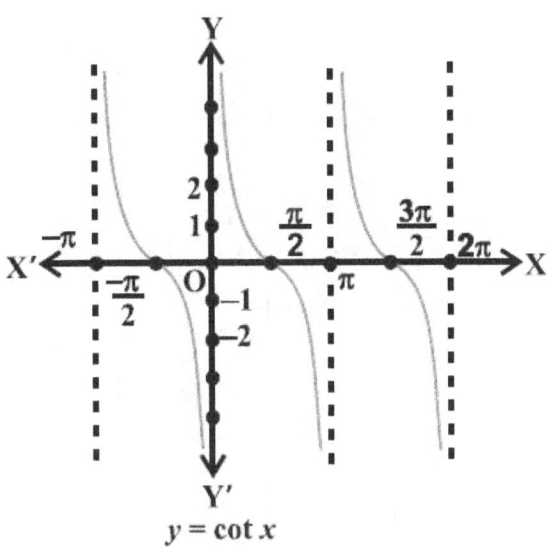

$$y = \cot x$$

Domain and Principal Branch of Inverse Cosecant

For inverse of Cosecant function, the domain is positive in first quadrant and negative in fourth quadrant.

The principal branch for Cosecant is first and fourth quadrant.

If $cosec^{-1} y = x$, then

$$0 < x \le \frac{\pi}{2} \text{ , for } y \ge 1$$

$$\text{and } -\frac{\pi}{2} \le x < 0 \text{, for } y \le -1$$

For function $cosec^{-1} y = x$, the principal branch is $[-\frac{\pi}{2} , 0)$

and $(0 , \dfrac{\pi}{2}]$; Domain is $y \geq 1$ and $y \leq -1$.

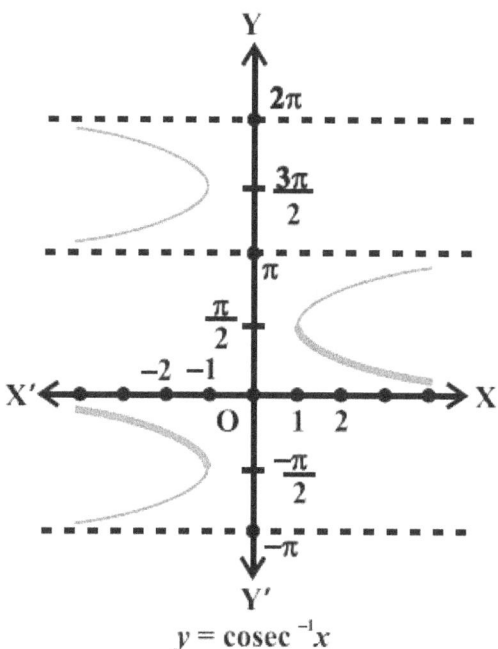

$$y = \text{cosec}^{-1}x$$

Domain/Range of Cosecant:

For non-inverse function $y = \text{cosec } x$, if angle is not limited to principal branch, the domain (all possible values of x) is the set $\{ x :$ $x \in \mathbf{R}$ and $x \neq n\pi,\ n \in \mathbf{Z} \}$ and range (all possible values of y) is the set $\{ y : y \in \mathbf{R},\ y \geq 1 \text{ or } y \leq -1 \}$

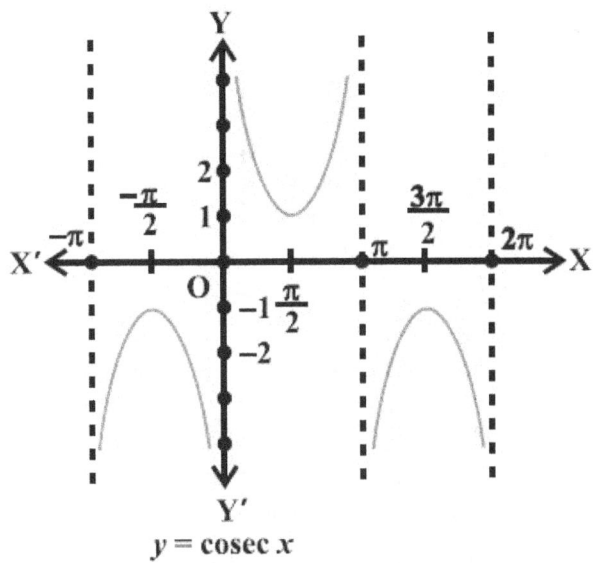

$y = \operatorname{cosec} x$

Summary of domain/principal branch of inverse functions:

Function name	Function	Domain	Pr. Branch
inverse Sine	$\sin^{-1} x$	$[-1, 1]$	$[-\frac{\pi}{2}, \frac{\pi}{2}]$
inverse Cosine	$\cos^{-1} x$	$[-1, 1]$	$[0, \pi]$
inverse Tangent	$\tan^{-1} x$	$(-\infty, \infty)$	$(-\frac{\pi}{2}, \frac{\pi}{2})$
inverse Cotangent	$\cot^{-1} x$	$(-\infty, \infty)$	$(0, \pi)$

| inverse Secant | $\sec^{-1} x$ | $(-\infty, -1]$ and $[1, \infty)$ | $[0, \frac{\pi}{2})$ and $(\frac{\pi}{2}, \pi]$ |

| inverse Cosecant | $\mathbf{cosec}^{-1} x$ | $(-\infty, -1]$ and $[1,\infty)$ | $[-\frac{\pi}{2}, 0)$ and $(0, \frac{\pi}{2}]$ |

6.4 Forward Inverse Identities

A forward-inverse trigonometric function is a function of the form f(g(x)), where f(x) is a non-inverse known as forward trigonometric function; and g(x) is an inverse trigonometric function. Examples are $\cos(\sin^{-1}(x))$, $\tan(\sin^{-1}(x))$ etc.

A forward-inverse trigonometric function shows the relationships between inverse and non-inverse trigonometric functions.

To evaluate any forward trigonometric function with inverse of sin, we select a triangle whose inverse of sin is x, ($x \in [-\frac{\pi}{2}, \frac{\pi}{2}]$) as depicted below:

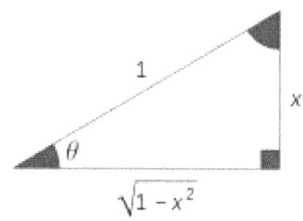

Inv-6.4.1: $\sin(\sin^{-1}(x)) = x$

Inv-6.4.2: $\cos(\sin^{-1}(x)) = \sqrt{1 - x^2}$

Inv-6.4.3: $\tan(\sin^{-1}(x)) = \dfrac{x}{\sqrt{1 - x^2}}$

To evaluate any forward trigonometric function with inverse of cos, we select a triangle whose inverse of cos is x, ($x \in [0, \pi]$) as depicted below:

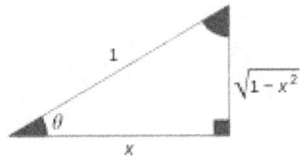

Inv-6.4.4: $\sin(\cos^{-1}(x)) = \sqrt{1 - x^2}$

Inv-6.4.5: $\cos(\cos^{-1}(\beta)) = x$

Inv-6.4.6: $\tan(\cos^{-1}(x)) = \dfrac{\sqrt{1-x^2}}{x}$

To evaluate any forward trigonometric function with inverse of tan, we select a triangle whose inverse of tan is x, ($x \in (-\dfrac{\pi}{2}, \dfrac{\pi}{2})$) as depicted below:

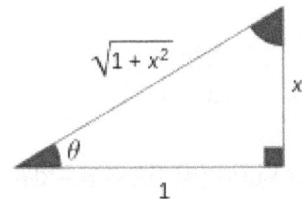

Inv-6.4.7: $\sin(\tan^{-1}(x)) = \dfrac{x}{\sqrt{1+x^2}}$

Inv-6.4.8: $\cos(\tan^{-1}(x)) = \dfrac{1}{\sqrt{1+x^2}}$

Inv-6.4.9: $\tan(\tan^{-1}(x)) = x$

$$\cos(\sin^{-1} x) = \sqrt{1 - x^2} \quad \sin(\cos^{-1} x) = \sqrt{1 - x^2}$$

$$\cos(\tan^{-1} x) = \frac{1}{\sqrt{1 + x^2}} \quad \sin(\tan^{-1} x) = \frac{x}{\sqrt{1 + x^2}}$$

$$\tan(\cos^{-1} x) = \frac{\sqrt{1 - x^2}}{x} \quad \tan(\sin^{-1} x) = \frac{x}{\sqrt{1 - x^2}}$$

6.5 Inverse Forward Identities

An inverse-forward trigonometric function is a function of the form g(f(x)), where g(x) is an inverse trigonometric function; and f(x) is a non-inverse (also called forward) trigonometric function. Examples are $\cos(\sin^{-1}(x))$, $\tan(\sin^{-1}(x))$ etc.

All of the non-inverse identities mentioned in previous sections, we can transform them into inverse forward identities, if the principal branch for inverse function is maintained appropriately.

For non-inverse function, one of the complementary identities (also called cofunction identities) is: $\sin\left(\frac{\pi}{2} - x\right) = \cos x$

We can transform it into inverse-forward identity as:

$$\sin^{-1}(\cos x) = \frac{\pi}{2} - x$$

Since the principal branch of inverse Sine is $[-\frac{\pi}{2}, \frac{\pi}{2}]$, so the value of $\sin^{-1}(\cos x)$ which is $\left(\frac{\pi}{2} - x\right)$ should have the set of values in [0, π] to satisfy the principal branch of Sine.

Inverse forward identities are as follows:-

Inv-6.5.1: $\sin^{-1}(\cos x) = \dfrac{\pi}{2} - x$, for $x \in [0, \pi]$

Inv-6.5.2: $\cos^{-1}(\sin x) = \dfrac{\pi}{2} - x$, for $x \in [-\dfrac{\pi}{2}, \dfrac{\pi}{2}]$

Inv-6.5.3: $\tan^{-1}(\cot x) = \dfrac{\pi}{2} - x$, for $x \in (0, \pi)$

Inv-6.5.4: $\cot^{-1}(\tan x) = \dfrac{\pi}{2} - x$, for $x \in (-\dfrac{\pi}{2}, \dfrac{\pi}{2})$

Inv-6.5.5: $\sec^{-1}(\operatorname{cosec} x) = \dfrac{\pi}{2} - x$, for $x \in (0, \dfrac{\pi}{2}]$ and $[-\dfrac{\pi}{2}, 0)$

Inv-6.5.6: $\operatorname{cosec}^{-1}(\sec x) = \dfrac{\pi}{2} - x$, for $x \in [0, \dfrac{\pi}{2})$ and $(\dfrac{\pi}{2}, \pi]$

Inverse-forward identities

$\sin^{-1}(\cos x) = \dfrac{\pi}{2} - x$, for $x \in [0, \pi]$

$\cos^{-1}(\sin x) = \dfrac{\pi}{2} - x$, for $x \in [-\dfrac{\pi}{2}, \dfrac{\pi}{2}]$

$\tan^{-1}(\cot x) = \dfrac{\pi}{2} - x$, for $x \in (0, \pi)$

$\cot^{-1}(\tan x) = \dfrac{\pi}{2} - x$, for $x \in (-\dfrac{\pi}{2}, \dfrac{\pi}{2})$

$\sec^{-1}(\operatorname{cosec} x) = \dfrac{\pi}{2} - x$, for $x \in (0, \dfrac{\pi}{2}]$ and $[-\dfrac{\pi}{2}, 0)$

$\operatorname{cosec}^{-1}(\sec x) = \dfrac{\pi}{2} - x$, for $x \in [0, \dfrac{\pi}{2})$ and $(\dfrac{\pi}{2}, \pi]$

Worksheet-15

Exercise-1: Write the equivalent inverse function for $\cos x = A$

Exercise-2: Write the equivalent non-inverse (forward) function for $sin^{-1} 1 = \frac{\pi}{2}$

Exercise-3: Write the equivalent non-inverse forward function for $tan^{-1} 1 = \frac{\pi}{4}$

Exercise-4: Evaluate $sin^{-1}(sin\frac{\pi}{4})$

Exercise-5: Evaluate $cos^{-1}(cos\frac{\pi}{4})$

Exercise-6: Evaluate $tan\,(tan^{-1}(-10))$

Exercise-7: Find principal value of $sin^{-1}\left(-\frac{\sqrt{3}}{2}\right)$

Exercise-8: Find principal value of $cos^{-1}(-\frac{1}{\sqrt{2}})$

Exercise-9: Evaluate the following:

$sin^{-1}(-1)$

$sin^{-1}(1)$

$tan^{-1}(0)$

$cos^{-1}(\frac{1}{\sqrt{2}})$

$cos^{-1}(-\frac{1}{\sqrt{2}})$

$cos^{-1}(\frac{1}{2})$

$$\cos^{-1}\left(-\frac{1}{2}\right)$$

$$\tan^{-1}(1)$$

$$\tan^{-1}(-1)$$

Exercise-10: Write the forward-inverse identities of following:

$\cos(\sin^{-1}(x)) =$

$\tan(\sin^{-1}(x)) =$

$\sin(\cos^{-1}(x)) =$

$\tan(\cos^{-1}(x)) =$

$\sin(\tan^{-1}(x)) =$

$\cos(\tan^{-1}(x)) =$

Exercise-11: Write the following inverse function in terms of other trigonometric functions.

$$\sin^{-1}(x) = \cos^{-1}(\underline{\quad}) = \tan^{-1}(\underline{\quad}) = \cot^{-1}(\underline{\quad})$$
$$= \sec^{-1}(\underline{\quad}) = \csc^{-1}(\underline{\quad})$$

$$\cos^{-1}(x) = \sin^{-1}(\underline{\quad}) = \tan^{-1}(\underline{\quad}) = \cot^{-1}(\underline{\quad})$$
$$= \sec^{-1}(\underline{\quad}) = \csc^{-1}(\underline{\quad})$$

$$\tan^{-1}(x) = \sin^{-1}(\underline{\quad}) = \cos^{-1}(\underline{\quad}) = \cot^{-1}(\underline{\quad})$$
$$= \sec^{-1}(\underline{\quad}) = \csc^{-1}(\underline{\quad})$$

Exercise-12: Fill in the blanks for inverse-forward identities of following:

$$\text{for } x \in [0, \pi], \quad \sin^{-1}(\underline{\quad}) = \frac{\pi}{2} - x$$

$$\text{for } x \in (0, \pi), \quad \tan^{-1}(\underline{\quad}) = \frac{\pi}{2} - x$$

$$\text{for } x \in (0, \frac{\pi}{2}] \text{ and } [-\frac{\pi}{2}, 0), \quad \sec^{-1}(\underline{\quad}) = \frac{\pi}{2} - x$$

Exercise-13: Write the principal branch and domain of inverse Sine.

Exercise-14: Write the principal branch and domain of inverse function of Cosine.

Exercise-15: Write the principal branch and domain of inverse function of Tangent.

Exercise-16: Write the principal branch and domain of inverse function of Cosecant.

Exercise-17: Write the principal branch and domain of inverse function of Secant.

Exercise-18: Write the principal branch and domain of inverse function of Cotangent.

Exercise-19: Solve for x

$$sin^{-1}(x + 1) = \frac{\pi}{3}$$

Exercise-20: Solve for x

$$tan^{-1}(x-1) = \frac{\pi}{4}$$

Exercise-21: Solve for x

$$\cos x^2 = -1$$

Exercise-22: Solve for x

$$\tan(x-2) = 1$$

Exercise-23: Evaluate $sin^{-1}(sin\frac{3\pi}{4})$

Exercise-24: Write the domain and primary branch in following table:

Function name	Function	Domain	Principal-Branch/ Range
inverse Sine	$\sin^{-1} x$		
inverse Cosine	$\cos^{-1} x$		
inverse Tangent	$\tan^{-1} x$		
inverse Cotangent	$\cot^{-1} x$		
inverse Secant	$\sec^{-1} x$		
inverse Cosecant	$\operatorname{cosec}^{-1} x$		

Solutions-Worksheet-15

Solution-1: The equivalent inverse function is $\cos^{-1} A = x$

Solution-2: The equivalent forward function is $\sin \frac{\pi}{2} = 1$

Solution-3: The equivalent forward function is $\tan \frac{\pi}{4} = 1$

Solution-4: As per inverse function relation, $\sin^{-1}(\sin y) = y$.

Therefore, $\sin^{-1}(\sin \frac{\pi}{4}) = \frac{\pi}{4}$

Alternatively, we can evaluate substituting the appropriate value, without using inverse relation.

$\sin^{-1}(\sin \frac{\pi}{4}) = \sin^{-1}(\frac{1}{\sqrt{2}}) = \frac{\pi}{4}$

Solution-5: As per inverse function relation, $\cos^{-1}(\cos y) = y$.

Therefore, $\cos^{-1}(\cos \frac{\pi}{4}) = \frac{\pi}{4}$

Solution-6: As per inverse function relation, $\tan (\tan^{-1} y) = y$.

Therefore, $\tan (\tan^{-1}(-10)) = -10$

Solution-7: In third and fourth quadrant, value of Sine is negative. Angles $-\frac{5\pi}{6}$ in third quadrant and $-\frac{\pi}{3}$ in fourth quadrant are the angles whose Sines produce the same value $-\frac{\sqrt{3}}{2}$.

But the angle of smallest absolute value is $-\frac{\pi}{3}$, which lies in the principal branch of inverse Sine is $[-\frac{\pi}{2}, \frac{\pi}{2}]$. Therefore only $-\frac{\pi}{3}$ is the principal value.

Solution-8: The corresponding acute angle for value $\frac{1}{\sqrt{2}}$ of Cosine is $\frac{\pi}{4}$.

As the value is negative, Cosine must be in second or third quadrant. Principal branch (range) of Cosine is $[0, \pi]$. So we must select the angle in second quadrant. Second quadrant angle whose corresponding acute angle is $\frac{\pi}{4}$ must be $(\pi - \frac{\pi}{4}) = \frac{3\pi}{4}$, which lies in the principal branch (range) of Cosine $[0, \pi]$.

Solution-9:

$sin^{-1}(-1) = -\frac{\pi}{2}$; The principal branch of inverse Sine is $[-\frac{\pi}{2}$, $\frac{\pi}{2}]$. Therefore only $-\frac{\pi}{2}$ is the principal value.

$sin^{-1}(1) = \frac{\pi}{2}$

$tan^{-1}(0) = 0$; Principal branch is $(-\frac{\pi}{2}, \frac{\pi}{2})$

$cos^{-1}(\frac{1}{\sqrt{2}}) = \frac{\pi}{4}$; Principal branch is $[0, \pi]$

$cos^{-1}(-\frac{1}{\sqrt{2}}) = \frac{3\pi}{4}$; The corresponding acute angle for value $\frac{1}{\sqrt{2}}$ of Cosine is $\frac{\pi}{4}$. As the value is negative, Cosine must be in second quadrant, because principal branch of Cosine is $[0, \pi]$. Second quadrant angle whose corresponding acute angle is $\frac{\pi}{4}$ must be $(\pi - \frac{\pi}{4}) = \frac{3\pi}{4}$

$cos^{-1}(\frac{1}{2}) = \frac{\pi}{3}$, which lies in the principal branch $[0, \pi]$

$cos^{-1}\left(-\frac{1}{2}\right) = (\pi - \frac{\pi}{3}) = \frac{2\pi}{3}$, which lies in the principal branch **[0, π]**

$tan^{-1}(1) = \frac{\pi}{4}$, which lies in principal branch $\left(-\frac{\pi}{2}, \frac{\pi}{2}\right)$

$tan^{-1}(-1) = -\frac{\pi}{4}$; The corresponding acute angle for value 1 of Tangent is $\frac{\pi}{4}$. As the value is negative, Tangent must be in fourth quadrant, because principal branch of Tangent is $\left(-\frac{\pi}{2}, \frac{\pi}{2}\right)$. Fourth quadrant angle whose corresponding acute angle is $\frac{\pi}{4}$ must be $\left(-\frac{\pi}{4}\right)$

Solution-10:

$\cos(\sin^{-1}(x)) = \sqrt{1-x^2}$; $\tan(\sin^{-1}(x)) = \frac{x}{\sqrt{1-x^2}}$;

$\sin(\cos^{-1}(x)) = \sqrt{1-x^2}$; $\tan(\cos^{-1}(x)) = \frac{\sqrt{1-x^2}}{x}$;

$\sin(\tan^{-1}(x)) = \frac{x}{\sqrt{1+x^2}}$; $\cos(\tan^{-1}(x)) = \frac{1}{\sqrt{1+x^2}}$;

Solution-11:

$\sin^{-1}(x) = \cos^{-1}(\sqrt{1-x^2}) = \tan^{-1}\left(\frac{x}{\sqrt{1-x^2}}\right) = \cot^{-1}\left(\frac{\sqrt{1-x^2}}{x}\right)$

$= \sec^{-1}\left(\frac{1}{\sqrt{1-x^2}}\right) = \csc^{-1}\left(\frac{1}{x}\right)$

$$\cos^{-1}(x) = \sin^{-1}(\sqrt{1-x^2}) = \tan^{-1}\left(\frac{\sqrt{1-x^2}}{x}\right) = \cot^{-1}\left(\frac{x}{\sqrt{1-x^2}}\right)$$

$$= \sec^{-1}\left(\frac{1}{x}\right) = \csc^{-1}\left(\frac{1}{\sqrt{1-x^2}}\right)$$

$$\tan^{-1}(x) = \sin^{-1}\left(\frac{x}{\sqrt{1+x^2}}\right) = \cos^{-1}\left(\frac{1}{\sqrt{1+x^2}}\right) = \cot^{-1}\left(\frac{1}{x}\right)$$

$$= \sec^{-1}(\sqrt{1+x^2}) = \csc^{-1}\left(\frac{\sqrt{1+x^2}}{x}\right)$$

Solution-12:

$$\sin^{-1}(\cos x) = \frac{\pi}{2} - x \text{, for } x \in [0, \pi]$$

$$\tan^{-1}(\cot x) = \frac{\pi}{2} - x \text{, for } x \in (0, \pi)$$

$$\sec^{-1}(\csc x) = \frac{\pi}{2} - x \text{, for } x \in (0, \frac{\pi}{2}] \text{ and } [-\frac{\pi}{2}, 0)$$

Solution-13: Inverse Sine Function

Principal Branch: $[-\frac{\pi}{2}, \frac{\pi}{2}]$;   Domain: $[-1, 1]$

Solution-14: Inverse Cosine Function

Principal Branch: $[0, \pi]$; Domain: $[-1, 1]$

Solution-15: Inverse Tangent Function

Principal Branch: $(-\frac{\pi}{2}, \frac{\pi}{2})$; Domain: $(-\infty, \infty)$

Solution-16: Inverse Cosecant Function

Principal Branch: $[-\frac{\pi}{2}, 0)$ **and** $(0, \frac{\pi}{2}]$; Domain: $(-\infty, -1]$ **and** $[1, \infty)$

Solution-17: Inverse Secant Function

Principal Branch: $[0, \frac{\pi}{2})$ **and** $(\frac{\pi}{2}, \pi]$; Domain: $(-\infty, -1]$ and $[1, \infty)$

Solution-18: Inverse Cotangent Function

Principal Branch: $(0, \pi)$; Domain: $(-\infty, \infty)$

Solution-19: As per definition of inverse function,

$$x + 1 = sin\,\frac{\pi}{3}$$

$$x = \frac{\sqrt{3}}{2} - 1 = \frac{\sqrt{3} - 2}{2} = \frac{-0.268}{2} = -0.134$$

Solution-20: As per definition of inverse function,

$$x - 1 = tan\,\frac{\pi}{4}$$

$$x = 1 + 1 = 2$$

Solution-21:

$$x^2 = cos^{-1}(-1) = \pi$$

$$x = \pm\sqrt{\pi}$$

Solution-22:

$$x - 2 = tan^{-1}(1)$$

$$x = 2 + \frac{\pi}{4}$$

Solution-23: As per inverse function relation, $sin^{-1}(sin\,y) = y$.

$sin^{-1}(sin\frac{3\pi}{4}) = \frac{3\pi}{4}$; But the principal branch of inverse Sine is $[-\frac{\pi}{2}, \frac{\pi}{2}]$ and $\frac{3\pi}{4} \notin [-\frac{\pi}{2}, \frac{\pi}{2}]$, so we will transform the second quadrant angle into first quadrant angle using supplementary identity.

$$sin\frac{3\pi}{4} = sin\left(\pi - \frac{3\pi}{4}\right) = sin\left(\frac{\pi}{4}\right)$$

$\frac{\pi}{4}$ is the principal value, because $\frac{\pi}{4} \in [-\frac{\pi}{2}, \frac{\pi}{2}]$, therefore $\frac{\pi}{4}$ is the solution.

Solution-24:

Function	Domain	Pr. Branch
$\sin^{-1} x$	$[-1, 1]$	$[-\frac{\pi}{2}, \frac{\pi}{2}]$
$\cos^{-1} x$	$[-1, 1]$	$[0, \pi]$
$\tan^{-1} x$	$(-\infty, \infty)$	$(-\frac{\pi}{2}, \frac{\pi}{2})$
$\cot^{-1} x$	$(-\infty, \infty)$	$(0, \pi)$
$\sec^{-1} x$	$(-\infty, -1]$ and $[1, \infty)$	$[0, \frac{\pi}{2})$ and $(\frac{\pi}{2}, \pi]$
$\operatorname{cosec}^{-1} x$	$(-\infty, -1]$ and $[1, \infty)$	$[-\frac{\pi}{2}, 0)$ and $(0, \frac{\pi}{2}]$

6.6   Conventional Inverse Identities

To prove/verify the formulas and identities of inverse functions, the inverse expression is converted into non-inverse forward expression while maintaining the condition of principal value branch.

6.6.1   Reciprocal Inverse Identities

Reciprocal Identities of Inverse

Inv-6.6.1: $\sin^{-1}\dfrac{1}{\beta} = \csc^{-1}\beta$, where $\beta \geq 1$ or $\beta \leq -1$

Let's assume $\csc^{-1}\beta = y$, i.e., $\beta = \csc y$

So, $\dfrac{1}{\beta} = \sin y$

Hence, $\sin^{-1}\dfrac{1}{\beta} = y$

Substituting the value of y, we have

$$\sin^{-1}\frac{1}{\beta} = \csc^{-1}\beta$$

Remark: Note the condition. As $\csc^{-1}\beta$ *has the domain* $(-\infty, -1]$ **and** $[1,\infty)$, *so* $\beta \geq 1$ *or* $\beta \leq -1$

Similarly,

Inv-6.6.1.B, $\csc^{-1}\dfrac{1}{\beta} = \sin^{-1}\beta$; where $\beta \geq -1$ or $\beta \leq 1$

Inv-6.6.2: $\cos^{-1}\dfrac{1}{\beta} = \sec^{-1}\beta$, where $\beta \geq 1$ or $\beta \leq -1$

Let's assume $\sec^{-1}\beta = y$, i.e., $\beta = \sec y$

So, $\dfrac{1}{\beta} = \cos y$

Hence, $\cos^{-1}\dfrac{1}{\beta} = y$

Substituting the value of y, we have

$$\cos^{-1}\dfrac{1}{\beta} = \sec^{-1}\beta$$

Remark: Note the condition. As $\sec^{-1}\beta$ *has the domain* (–∞, –1] **and** [1, ∞), *so* $\beta \geq 1$ *or* $\beta \leq -1$

Similarly,

Inv-6.6.2.B: $\sec^{-1}\dfrac{1}{\beta} = \cos^{-1}\beta$, where $\beta \geq -1$ or $\beta \leq 1$

Inv-6.6.3.A1: $\tan^{-1}\dfrac{1}{\beta} = \cot^{-1}\beta$; where $\beta > 0$

Let's assume $\cot^{-1}\beta = y$, i.e., $\beta = \cot y$

So, $\dfrac{1}{\beta} = \tan y$

Hence, $\tan^{-1}\dfrac{1}{\beta} = y$

Substituting the value of y, we have

$$\tan^{-1}\dfrac{1}{\beta} = \cot^{-1}\beta$$

Alternatively, we can also derive complementary reciprocal identity:

Inv-6.6.3-A2: $\quad \tan^{-1} \dfrac{1}{\beta} = \dfrac{\pi}{2} - \tan^{-1} \beta$, if $\beta > 0$

The complementary identity of inverse is:

$$\tan^{-1} \beta + \cot^{-1} \beta = \dfrac{\pi}{2}$$

Simplifying this equation, we have

$$\cot^{-1} \beta = \dfrac{\pi}{2} - \tan^{-1} \beta$$

In **Inv-6.6.3**, we mentioned that $\tan^{-1} \dfrac{1}{\beta} = \cot^{-1} \beta$, so

$$\tan^{-1} \dfrac{1}{\beta} = \dfrac{\pi}{2} - \tan^{-1} \beta$$

Therefore, the complete reciprocal identity for inverse tan, if $\beta > 0$ is stated as:

Inv-6.6.3-A: $\quad \tan^{-1} \dfrac{1}{\beta} = \dfrac{\pi}{2} - \tan^{-1} \beta = \cot^{-1} \beta$, where $\beta > 0$

Remark: Note that as $\tan^{-1} x$ has the principal branch as $\left(-\dfrac{\pi}{2}, \dfrac{\pi}{2}\right.$), so negative domain is only possible in fourth quadrant in principal branch $\left(-\dfrac{\pi}{2}, 0\right)$.

But $\cot^{-1} x$ has the principal branch as $(0, \pi)$, so negative domain is only possible in second quadrant in principal branch $\left(\dfrac{\pi}{2}, \pi\right)$.

So, if β < 0 then the equation $\tan^{-1}\dfrac{1}{\beta} = \cot^{-1}\beta$ *is invalid. The valid reciprocal identity of inverse tan, if β<0 is mentioned below:*

Inv-6.6.3-B1: $\tan^{-1}\dfrac{1}{\beta} = \cot^{-1}\beta - \pi$, where $\beta < 0$

Applying supplementary identity $\cot(\pi - \beta) = -\cot\beta$ with **Inv-6.6.3.A1**, we can prove **Inv-6.6.3-B1.**

Similarly the complementary angle identity is:

Inv-6.6.3-B2: $\tan^{-1}\dfrac{1}{\beta} = \dfrac{-\pi}{2} - \tan^{-1}\beta$, where $\beta < 0$

Therefore, the complete reciprocal identity for inverse tan, if β < 0 is stated as:

Inv-6.6.3-B: $\tan^{-1}\dfrac{1}{\beta} = \dfrac{-\pi}{2} - \tan^{-1}\beta = \cot^{-1}\beta - \pi$, where $\beta < 0$

Similarly the reciprocal identities of inverse cot are:

Inv-6.6.3-C1: $\cot^{-1}\dfrac{1}{\beta} = \tan^{-1}\beta$, for $\beta > 0$

Inv-6.6.3-C2: $\cot^{-1}\dfrac{1}{\beta} = \dfrac{\pi}{2} - \cot^{-1}\beta$, for $\beta > 0$

Therefore, the complete reciprocal identity for inverse tan, if β > 0 is stated as:

Inv-6.6.3-C: $\cot^{-1}\dfrac{1}{\beta} = \dfrac{\pi}{2} - \cot^{-1}\beta = \tan^{-1}\beta$, for $\beta > 0$

Inv-6.6.3-D1: $\cot^{-1}\dfrac{1}{\beta} = \tan^{-1}\beta + \pi$, for $\beta < 0$

Inv-6.6.3-D2: $\cot^{-1}\dfrac{1}{\beta} = \dfrac{3\pi}{2} - \cot^{-1}\beta$, for $\beta < 0$

Therefore, the complete reciprocal identity for inverse tan, if $\beta < 0$ is stated as:

Inv-6.6.3-D: $\cot^{-1}\dfrac{1}{\beta} = \dfrac{3\pi}{2} - \cot^{-1}\beta = \tan^{-1}\beta + \pi$, for $\beta < 0$

Reciprocal Identities of Inverse

Inv-6.6.1: $\sin^{-1}\dfrac{1}{\beta} = \operatorname{cosec}^{-1}\beta$, where $\beta \geq 1$ and $\beta \leq -1$

Inv-6.6.1.B: $\operatorname{cosec}^{-1}\dfrac{1}{\beta} = \sin^{-1}\beta$, where $\beta \geq -1$ and $\beta \leq 1$

Inv-6.6.2: $\cos^{-1}\dfrac{1}{\beta} = \sec^{-1}\beta$, where $\beta \geq 1$ and $\beta \leq -1$

Inv-6.6.2.B: $\sec^{-1}\dfrac{1}{\beta} = \cos^{-1}\beta$, where $\beta \geq -1$ and $\beta \leq 1$

Inv-6.6.3-A: $\tan^{-1}\dfrac{1}{\beta} = \dfrac{\pi}{2} - \tan^{-1}\beta = \cot^{-1}\beta$, where $\beta > 0$

Inv-6.6.3-B: $\tan^{-1}\dfrac{1}{\beta} = \dfrac{-\pi}{2} - \tan^{-1}\beta = \cot^{-1}\beta - \pi$, where $\beta < 0$

Inv-6.6.3-C: $\cot^{-1}\dfrac{1}{\beta} = \dfrac{\pi}{2} - \cot^{-1}\beta = \tan^{-1}\beta$, for $\beta > 0$

Inv-6.6.3-D: $\cot^{-1}\dfrac{1}{\beta} = \dfrac{3\pi}{2} - \cot^{-1}\beta = \tan^{-1}\beta + \pi$, for $\beta < 0$

6.6.2   Negative Inverse Identities

Inv-6.6.4: $\sin^{-1}(-\beta) = -\sin^{-1}\beta$; where $\beta \in [-1, 1]$

Note that $\sin^{-1}\beta$ has the domain [-1, 1]

Let's assume $\sin^{-1}(-\beta) = y$, i.e., $-\beta = \sin y$, so

$\beta = -\sin y$

Applying the negative identity of sin, we have

$\beta = \sin(-y)$,

So, $\sin^{-1}\beta = -y$. Substituting y, we get

$\sin^{-1}\beta = -\sin^{-1}(-\beta)$

Hence, $\sin^{-1}(-\beta) = -\sin^{-1}\beta$

Inv-6.6.5: $\csc^{-1}(-\beta) = -\csc^{-1}\beta$; where $|\beta| \geq 1$

Note that $\csc^{-1}\beta$ has the domain $(-\infty, -1]$ and $[1, \infty)$. So $|\beta| \geq 1$.

Using negative identity of cosec, we can prove **Inv-6.6.5** similar to **Inv-6.6.4**

Inv-6.6.6: $\tan^{-1}(-\beta) = -\tan^{-1}\beta$; where $\beta \in \mathbf{R}$

Note that $\tan^{-1}\beta$ has the domain $(-\infty, \infty)$, i.e. $\beta \in \mathbf{R}$

Using negative identity of tan, we can prove **Inv-6.6.6** similar to **Inv-6.6.4**

Negative Identities of Inverse

Inv-6.6.4: $\quad \sin^{-1}(-\beta) = -\sin^{-1}\beta$; where $\beta \in [-1, 1]$

Inv-6.6.5: $\quad \csc^{-1}(-\beta) = -\csc^{-1}\beta$; where $|\beta| \geq 1$

Inv-6.6.6: $\quad \tan^{-1}(-\beta) = -\tan^{-1}\beta$; where $\beta \in \mathbf{R}$

6.6.3 Negative Supplementary Identities of Inverse

Inv-6.6.7: $\quad \cos^{-1}(-\beta) = \pi - \cos^{-1}\beta$; where $\beta \in [-1, 1]$

Note that $\cos^{-1}\beta$ has the domain [-1, 1]

Let's assume $\cos^{-1}(-\beta) = y$, i.e., $-\beta = \cos y$, so using supplementary identity of cos, we get

$$\beta = -\cos y = \cos(\pi - y)$$

Applying inverse of cos at both sides of equation $\beta = \cos(\pi - y)$, we have,

$$\cos^{-1}\beta = \pi - y$$

Substituting for y, we get

$$\cos^{-1}\beta = \pi - \cos^{-1}(-\beta)$$

Rearranging the equation, we get

$$\cos^{-1}(-\beta) = \pi - \cos^{-1}\beta$$

Inv-6.6.8: $\sec^{-1}(-\beta) = \pi - \sec^{-1}\beta$; where $|\beta| \geq 1$

Note that $\sec^{-1}\beta$ has the domain $(-\infty, -1]$ and $[1, \infty)$. So $|\beta| \geq 1$.

Using supplementary identity $\sec(\pi - y) = -\sec y$, we can prove **Inv-6.6.8** similar to **Inv-6.6.7**

Inv-6.6.9: $\cot^{-1}(-\beta) = \pi - \cot^{-1}\beta$; where $\beta \in \mathbf{R}$

Note that $\cot^{-1}\beta$ has the domain $(-\infty, \infty)$, i.e. $\beta \in \mathbf{R}$

Using supplementary identity $\cot(\pi - y) = -\cot y$, we can prove **Inv-6.6.9** similar to **Inv-6.6.7**

Negative Supplementary Identities of Inverse
Inv-6.6.7: $\cos^{-1}(-\beta) = \pi - \cos^{-1}\beta$; where $\beta \in [-1, 1]$
Inv-6.6.8: $\sec^{-1}(-\beta) = \pi - \sec^{-1}\beta$; where $|\beta| \geq 1$
Inv-6.6.9: $\cot^{-1}(-\beta) = \pi - \cot^{-1}\beta$; where $\beta \in \mathbf{R}$

6.6.4   Complementary Inverse Identities

Inv-6.6.10: $\sin^{-1}\beta + \cos^{-1}\beta = \dfrac{\pi}{2}$; where $\beta \in [-1, 1]$

Note that $\sin^{-1}\beta$ and $\cos^{-1}\beta$ have the domain $[-1, 1]$

Let's assume $\sin^{-1}\beta = y$, i.e., $\beta = \sin y$.

Applying the complimentary identity, we have

$$\beta = \sin y = \cos\left(\frac{\pi}{2} - y\right)$$

Therefore,

$$\cos^{-1}\beta = \frac{\pi}{2} - y = \frac{\pi}{2} - \sin^{-1}\beta$$

Hence, $\sin^{-1}\beta + \cos^{-1}\beta = \frac{\pi}{2}$

Inv-6.6.11: $\tan^{-1}\beta + \cot^{-1}\beta = \frac{\pi}{2}$; where $\beta \in \mathbf{R}$

Note that $\tan^{-1}\beta$ and $\cot^{-1}\beta$ have the domain $(-\infty, \infty)$, i.e. $\beta \in \mathbf{R}$

Using the complimentary identity $\cot\left(\frac{\pi}{2} - y\right) = \tan y$, we can prove **Inv-6.6.11** similar to **Inv-6.6.10**.

Inv-6.6.12: $\operatorname{cosec}^{-1}\beta + \sec^{-1}\beta = \frac{\pi}{2}$; where $|\beta| \geq 1$

Note that $\operatorname{cosec}^{-1}\beta$ and $\sec^{-1}\beta$ have the domain $(-\infty, -1]$ and $[1, \infty)$. So $|\beta| \geq 1$.

Using the complimentary identity $\sec\left(\frac{\pi}{2} - y\right) = \operatorname{cosec} y$, we can prove **Inv-6.6.12** similar to **Inv-6.6.10**.

Complementary Identities of Inverse

Inv-6.6.10: $\sin^{-1}\beta + \cos^{-1}\beta = \dfrac{\pi}{2}$; where $\beta \in [-1, 1]$

Inv-6.6.11: $\tan^{-1}\beta + \cot^{-1}\beta = \dfrac{\pi}{2}$; where $\beta \in \mathbf{R}$

Inv-6.6.12: $\operatorname{cosec}^{-1}\beta + \sec^{-1}\beta = \dfrac{\pi}{2}$; where $|\beta| \geq 1$

6.6.5   Sum/Difference Identities of Inverse

Inv-6.6.13: $\tan^{-1}\alpha + \tan^{-1}\beta = \tan^{-1}\dfrac{\alpha + \beta}{1 - \alpha\beta}$; where $\alpha\beta < 1$

Let's assume $\tan^{-1}\alpha = x,\ \tan^{-1}\beta = y$, then

$$\alpha = \tan x \quad \text{and} \quad \beta = \tan y$$

Applying sum identity of tan, we have

$$\tan(x + y) = \frac{\tan x + \tan y}{1 - \tan x\,\tan y} = \frac{\alpha + \beta}{1 - \alpha\beta}$$

So, $x + y = \tan^{-1}\left(\dfrac{\alpha + \beta}{1 - \alpha\beta}\right)$

Substituting x and y, we get

$$\tan^{-1}\alpha + \tan^{-1}\beta = \tan^{-1}\left(\frac{\alpha + \beta}{1 - \alpha\beta}\right)$$

Inv-6.6.14: $\tan^{-1}\alpha - \tan^{-1}\beta = \tan^{-1}\dfrac{\alpha - \beta}{1 + \alpha\beta}$; where $\alpha\beta > -1$

Applying difference identity of tan,

$$\tan (x - y) = \frac{\tan x - \tan y}{1 + \tan x \ \tan y} \ , \text{ we can prove } \textbf{Inv-6.6.14} \text{ similar}$$

to **Inv-6.6.13**.

Summarizing together the inverse sum/difference trigonometric identities of inverse, we have:

Inv-6.6.14-A: $\tan^{-1} \alpha \pm \tan^{-1} \beta = \tan^{-1} \left(\dfrac{\alpha \pm \beta}{1 \mp \alpha\beta} \right)$

Inv-6.6.14-B: $\sin^{-1} \alpha \pm \sin^{-1} \beta = \sin^{-1} \left(\alpha \sqrt{(1 - \beta^2)} \pm \beta \sqrt{(1 - \alpha^2)} \right)$

Inv-6.6.14-C: $\cos^{-1} \alpha \pm \cos^{-1} \beta = \cos^{-1} \left(\alpha\beta \mp \sqrt{(1 - \alpha^2)(1 - \beta^2)} \right)$

Inv-6.6.14-D: $\cot^{-1} \alpha \pm \cot^{-1} \beta = \cot^{-1} \left(\dfrac{\alpha\beta \mp 1}{\beta \pm \alpha} \right)$

Sum/Difference Identities of Inverse

Inv-6.6.14-A: $\tan^{-1} \alpha \pm \tan^{-1} \beta = \tan^{-1} \left(\dfrac{\alpha \pm \beta}{1 \mp \alpha\beta} \right)$

Inv-6.6.14-B: $\sin^{-1}\alpha \pm \sin^{-1}\beta = \sin^{-1}\left(\alpha \sqrt{(1 - \beta^2)} \pm \beta \sqrt{(1 - \alpha^2)} \right)$

Inv-6.6.14-C: $\cos^{-1}\alpha \pm \cos^{-1}\beta = \cos^{-1}\left(\alpha\beta \mp \sqrt{(1 - \alpha^2)(1 - \beta^2)} \right)$

Inv-6.6.14-D: $\cot^{-1} \alpha \pm \cot^{-1} \beta = \cot^{-1} \left(\dfrac{\alpha\beta \mp 1}{\beta \pm \alpha} \right)$

6.6.6   Double Angle Identities of Inverse

Inv-6.6.15: $2 \tan^{-1} \alpha = \sin^{-1} \dfrac{2\alpha}{1+\alpha^2}$, where $|\alpha| \leq 1$

Let's assume $\tan^{-1} \alpha = y$, then $\alpha = \tan y$

Considering RHS expression of equation,

$$\sin^{-1}\left(\frac{2\alpha}{1+\alpha^2}\right) = \sin^{-1}\left(\frac{2\tan y}{1+\tan^2 y}\right)$$

Applying the double angle identity $\sin 2\alpha = \dfrac{2\tan\alpha}{1+\tan^2\alpha}$, we have

$$\sin^{-1}\left(\frac{2\tan y}{1+\tan^2 y}\right) = \sin^{-1}(\sin 2y) = 2y$$

Substituting the value of y, we get $2y = 2\tan^{-1}\alpha$

Inv-6.6.16: $2\tan^{-1}\alpha = \cos^{-1}\dfrac{1-\alpha^2}{1+\alpha^2}$, where $\alpha \geq 1$

Applying the double angle identity $\cos 2\alpha = \dfrac{1-\tan^2\alpha}{1+\tan^2\alpha}$, we can prove **Inv-6.6.16** similar to **Inv-6.6.15**.

Inv-6.6.17: $2\tan^{-1}\alpha = \tan^{-1}\dfrac{2\alpha}{1-\alpha^2}$, where $-1 < \alpha < 1$

Applying the double angle identity $\tan 2\alpha = \dfrac{2\tan\alpha}{1-\tan^2\alpha}$, we can prove **Inv-6.6.17** similar to **Inv-6.6.15**.

Inv-6.6.18: $\sin^{-1}(2\alpha\sqrt{1-\alpha^2}) = 2\sin^{-1}\alpha$; where $\dfrac{-1}{\sqrt{2}} \leq \alpha \leq \dfrac{1}{\sqrt{2}}$

Let's assume $\sin^{-1}\alpha = y$, then $\alpha = \sin y$

Working at LHS of equation and substituting for α, we have

$$\sin^{-1}(2\,\alpha\,\sqrt{1-\alpha^2}) = \sin^{-1}(2\sin y\,\sqrt{1-\sin^2 y})$$

$$= \sin^{-1}(2\sin y\cos y)$$

Applying double angle identity $\sin 2y = 2\sin y\cos y$, we have

$$\sin^{-1}(2\sin y\cos y) = \sin^{-1}(\sin 2y) = 2y$$

Substituting back for y, we get: $2y = 2\sin^{-1}\alpha$

Double Angle Identities of Inverse

Inv-6.6.15: $2\tan^{-1}\alpha = \sin^{-1}\dfrac{2\alpha}{1+\alpha^2}$, where $|\alpha| \le 1$

Inv-6.6.16: $2\tan^{-1}\alpha = \cos^{-1}\dfrac{1-\alpha^2}{1+\alpha^2}$, where $\alpha \ge 1$

Inv-6.6.17: $2\tan^{-1}\alpha = \tan^{-1}\dfrac{2\alpha}{1-\alpha^2}$, where $-1 < \alpha < 1$

Inv-6.6.18: $\sin^{-1}(2\alpha\,\sqrt{1-\alpha^2}) = 2\sin^{-1}\alpha$; where $\dfrac{-1}{\sqrt{2}} \le \alpha \le \dfrac{1}{\sqrt{2}}$

Half Angle Identities of Inverse

Inv-6.6.19: $\sin^{-1}\alpha = 2\tan^{-1}\left(\dfrac{\alpha}{1+\sqrt{(1-\alpha^2)}}\right)$

Inv-6.6.20: $\cos^{-1}\alpha = 2\tan^{-1}\left(\dfrac{\sqrt{(1-\alpha^2)}}{1+\alpha}\right)$, if $-1 < \alpha \le 1$

Inv-6.6.21: $\tan^{-1}\alpha = 2\tan^{-1}\left(\dfrac{\alpha}{1+\sqrt{(1+\alpha^2)}}\right)$

Worksheet-16

Exercise-1: Find the value of $\sin^{-1}(\sin \frac{3\pi}{5})$

Exercise-2: Find the value of $\cos^{-1}(\cos -\frac{\pi}{4})$

Exercise-3: Find the value of $\cos^{-1}(\cos \pi)$

Exercise-4: Find the value of $\tan^{-1}(\tan \frac{3\pi}{5})$

Exercise-5: Find the value of $\csc^{-1}(\csc\frac{4\pi}{5})$

Exercise-6: Find the value of $\sec^{-1}(\sec\frac{7\pi}{5})$

Exercise-7: Find the value of $\sec^{-1}(\sec-\frac{\pi}{5})$

Exercise-8: Find the value of $\cot^{-1}(\cot \frac{11\pi}{3})$

Exercise-9: Find the value of $\cot^{-1}(\cot \frac{7\pi}{6})$

Exercise-10: Find the value of

$\cos(\sin^{-1}\beta + \cos^{-1}\beta)$; where $\beta \in [-1, 1]$

Exercise-11: Find the value of

$\csc (\tan^{-1} \beta + \cot^{-1} \beta)$; where $\beta \in \mathbf{R}$

Exercise-12: Find the value of

$\cot (\csc^{-1} \beta + \sec^{-1} \beta)$; where $| \beta| \geq 1$

Exercise-13: Find the value of $\cos^{-1} (-\frac{\sqrt{3}}{2})$

Exercise-14: Find the value of $\sec^{-1} (-\sqrt{2})$

Exercise-15: Find the value of $\cot^{-1}\left(-\frac{1}{\sqrt{3}}\right)$

Exercise-16: Verify $\cos^{-1}(2\alpha^2 - 1) = 2\cos^{-1}\alpha$

Exercise-17: Verify $\tan^{-1}\frac{1}{2} + \tan^{-1}\frac{2}{5} = \tan^{-1}\frac{9}{8}$

Exercise-18: Verify: $\tan^{-1}\frac{1}{2} - \tan^{-1}\frac{2}{5} = \tan^{-1}\left(\frac{1}{12}\right)$

Exercise-19: Write the reciprocal identities of inverse functions with the conditions.

Exercise-20: Write the negative identities of inverse functions with the conditions.

Exercise-21: Write the negative supplementary identities of inverse functions with the conditions.

Exercise-22: Write the complimentary identities of inverse functions with the conditions.

Exercise-23: Write the sum/difference identities of inverse functions with the conditions.

Exercise-24: Write the double angle identities of inverse functions with the conditions.

Solutions-Worksheet-16

Solution-1: As per inverse relation, $\sin^{-1}(\sin \beta) = \beta$. Therefore,

$$\sin^{-1}\left(\sin \frac{3\pi}{5}\right) = \frac{3\pi}{5}$$

Principal branch of inverse Sine is $[-\frac{\pi}{2}, \frac{\pi}{2}]$, but $\frac{3\pi}{5} \notin [-\frac{\pi}{2}, \frac{\pi}{2}]$

We apply supplementary identity to find principal value,

$$\sin\left(\frac{3\pi}{5}\right) = \sin\left(\pi - \frac{3\pi}{5}\right) = \sin\left(\frac{2\pi}{5}\right)$$

As $\frac{2\pi}{5} \in [-\frac{\pi}{2}, \frac{\pi}{2}]$, the value of $\sin^{-1}\left(\sin \frac{3\pi}{5}\right)$ is $\frac{2\pi}{5}$

Solution-2: As per inverse relation, $\cos^{-1}(\cos \beta) = \beta$. Therefore,

$$\cos^{-1}\left(\cos -\frac{\pi}{4}\right) = -\frac{\pi}{4}$$

Principal branch of inverse Cosine is $[0, \pi]$, but $-\frac{\pi}{4} \notin [0, \pi]$

We apply negative identity to find principal value,

$$\cos\left(-\frac{\pi}{4}\right) = \cos\frac{\pi}{4}$$

As $\frac{\pi}{4} \in [-\frac{\pi}{2}, \frac{\pi}{2}]$, the value of $\cos^{-1}\left(\cos -\frac{\pi}{4}\right)$ is $\frac{\pi}{4}$.

Solution-3: As per inverse relation, $\cos^{-1}(\cos \beta) = \beta$. Therefore,

$$\cos^{-1}(\cos \pi) = \pi$$

Principal branch of inverse Cosine is $[0, \pi]$, and $\pi \in [0, \pi]$

Therefore, the value of $\cos^{-1}(\cos \pi)$ is π.

Solution-4: As per inverse relation, $\tan^{-1}(\tan \beta) = \beta$. Therefore,

$$\tan^{-1}(\tan \frac{3\pi}{5}) = \frac{3\pi}{5}$$

Principal branch of inverse Cotangent is $(-\frac{\pi}{2}, \frac{\pi}{2})$, but $\frac{3\pi}{5} \notin (-\frac{\pi}{2}, \frac{\pi}{2})$

We apply supplementary and negative identities to find principal value,

$$\tan(\frac{3\pi}{5}) = \tan(\pi - \frac{3\pi}{5}) = -\tan(\frac{2\pi}{5}) = \tan(\frac{-2\pi}{5})$$

As $\frac{-2\pi}{5} \in [-\frac{\pi}{2}, \frac{\pi}{2}]$, the value of $\tan^{-1}(\tan \frac{3\pi}{5})$ is $\frac{-2\pi}{5}$

Solution-5: As per inverse relation, $\csc^{-1}(\csc \beta) = \beta$. Therefore,

$$\csc^{-1}(\csc \frac{4\pi}{5}) = \frac{4\pi}{5}$$

Principal branch of inverse Cosecant is $[-\frac{\pi}{2}, 0)$ and $(0, \frac{\pi}{2}]$, but $\frac{3\pi}{5} \notin [-\frac{\pi}{2}, 0)$ and $\frac{3\pi}{5} \notin (0, \frac{\pi}{2}]$

We apply supplementary identity to find principal value,

$$\csc(\frac{4\pi}{5}) = \csc(\pi - \frac{4\pi}{5}) = \csc(\frac{\pi}{5})$$

As $\frac{\pi}{5} \in (0, \frac{\pi}{2}]$, the value of $\csc^{-1}(\csc \frac{4\pi}{5})$ is $\frac{\pi}{5}$

Solution-6: As per inverse relation, $\sec^{-1}(\sec \beta) = \beta$. Therefore,

$$\sec^{-1}(\sec \frac{7\pi}{5}) = \frac{7\pi}{5}$$

Principal branch of inverse Secant is $[0, \frac{\pi}{2})$ **and** $(\frac{\pi}{2}, \pi]$, but $\frac{7\pi}{5} \notin$ $[0, \frac{\pi}{2})$ and $\frac{7\pi}{5} \notin (\frac{\pi}{2}, \pi]$

We apply $(2\pi - \beta)$ identity to find principal value,

$$\sec(\frac{7\pi}{5}) = \sec(2\pi - \frac{7\pi}{5}) = \sec(\frac{3\pi}{5})$$

As $\frac{3\pi}{5} \in (\frac{\pi}{2}, \pi]$, the value of $\sec^{-1}(\sec \frac{7\pi}{5})$ is $\frac{3\pi}{5}$

Solution-7: As per inverse relation, $\sec^{-1}(\sec \beta) = \beta$. Therefore,

$$\sec^{-1}(\sec -\frac{\pi}{5}) = -\frac{\pi}{5}$$

Principal branch of inverse Secant is $[0, \frac{\pi}{2})$ **and** $(\frac{\pi}{2}, \pi]$, but $\frac{-\pi}{5} \notin$ $[0, \frac{\pi}{2})$ and $\frac{-\pi}{5} \notin (\frac{\pi}{2}, \pi]$

We apply negative identity to find principal value,

$$\sec(-\frac{\pi}{5}) = \sec(\frac{\pi}{5})$$

As $\frac{\pi}{5} \in [0, \frac{\pi}{2})$, the value of $\sec^{-1}(\sec -\frac{\pi}{5})$ is $\frac{\pi}{5}$

Solution-8: As per inverse relation, $\cot^{-1}(\cot \beta) = \beta$. Therefore,

$$\cot^{-1}\left(\cot \frac{11\pi}{3}\right) = \frac{11\pi}{3}$$

Principal branch of inverse Cotangent is $(0, \pi)$, but $\frac{11\pi}{3} \notin (0, \pi)$

We apply $(2\pi - \beta)$ identity to find principal value,

$$\cot\left(\frac{11\pi}{3}\right) = \cot\left(2 \cdot 2\pi - \frac{\pi}{3}\right) = \cot\left(4\pi - \frac{\pi}{3}\right) = -\cot\left(\frac{\pi}{3}\right)$$

$$-\cot\left(\frac{\pi}{3}\right) = \cot\left(-\frac{\pi}{3}\right)$$

As $-\frac{\pi}{3} \notin (0, \pi)$, we apply supplementary identity to find principal value,

$$-\cot\left(\frac{\pi}{3}\right) = \cot\left(\pi - \frac{\pi}{3}\right) = \cot\left(\frac{2\pi}{3}\right)$$

As $\frac{2\pi}{3} \in (0, \pi)$, the value of $\cot^{-1}\left(\cot \frac{11\pi}{3}\right)$ is $\frac{2\pi}{3}$

Solution-9: As per inverse relation, $\cot^{-1}(\cot \beta) = \beta$. Therefore,

$$\cot^{-1}\left(\cot \frac{7\pi}{6}\right) = \frac{7\pi}{6}$$

Principal branch of inverse Cotangent is $(0, \pi)$, but $\frac{7\pi}{6} \notin (0, \pi)$

We apply $(\pi + \beta)$ identity to find principal value,

$$\cot\left(\frac{7\pi}{6}\right) = \cot\left(\pi + \frac{\pi}{6}\right) = \cot\left(\frac{\pi}{6}\right)$$

As $\frac{\pi}{6} \in (0, \pi)$, the value of $\cot^{-1}\left(\cot \frac{7\pi}{6}\right)$ is $\frac{\pi}{6}$

Solution-10: According to complementary identity, **Inv-6.6.10:**
$\sin^{-1}\beta + \cos^{-1}\beta = \dfrac{\pi}{2}$; where $\beta \in [-1, 1]$, so

$$\cos(\sin^{-1}\beta + \cos^{-1}\beta) = \cos\left(\dfrac{\pi}{2}\right) = 0$$

Solution-11: According to complementary identity, **Inv-6.6.11:**
$\tan^{-1}\beta + \cot^{-1}\beta = \dfrac{\pi}{2}$; where $\beta \in \mathbf{R}$, so

$$\operatorname{cosec}(\tan^{-1}\beta + \cot^{-1}\beta) = \operatorname{cosec}\left(\dfrac{\pi}{2}\right) = 1$$

Solution-12: According to complementary identity, **Inv-6.6.12:**
$\operatorname{cosec}^{-1}\beta + \sec^{-1}\beta = \dfrac{\pi}{2}$; where $|\beta| \geq 1$, so

$$\cot(\operatorname{cosec}^{-1}\beta + \sec^{-1}\beta) = \cot\left(\dfrac{\pi}{2}\right) = 0$$

Solution-13: Applying the supplementary identity of inverse **Inv-6.6.7** $\cos^{-1}(-\beta) = \pi - \cos^{-1}\beta$; where $\beta \in [-1, 1]$, we have

$$\cos^{-1}\left(-\dfrac{\sqrt{3}}{2}\right) = \pi - \cos^{-1}\left(\dfrac{\sqrt{3}}{2}\right) = \pi - \dfrac{\pi}{6} = \dfrac{5\pi}{6}$$

Note: Principal branch of inverse cos is $[0, \pi]$

Solution-14: Applying the supplementary identity of inverse **Inv-6.6.8** $\sec^{-1}(-\beta) = \pi - \sec^{-1}\beta$; where $|\beta| \geq 1$, we have

$$\sec^{-1}(-\sqrt{2}) = \pi - \sec^{-1}(\sqrt{2}) = \pi - \dfrac{\pi}{4} = \dfrac{3\pi}{4}$$

Note: Principal branch of inverse sec is $[0, \dfrac{\pi}{2})$ and $(\dfrac{\pi}{2}, \pi]$

Solution-15: applying the supplementary identity of inverse **Inv-6.6.9** $\cot^{-1}(-\beta) = \pi - \cot^{-1}\beta$; where $\beta \in \mathbf{R}$, we have

$$\cot^{-1}\left(-\frac{1}{\sqrt{3}}\right) = \pi - \cot^{-1}\left(\frac{1}{\sqrt{3}}\right) = \pi - \frac{\pi}{6} = \frac{5\pi}{6}$$

Note: Principal branch of inverse cot is **(0, π)**

Solution-16: Let's assume $\cos^{-1}\alpha = y$, then $\alpha = \cos y$

Working at LHS of equation and substituting for α, we have

$$\cos^{-1}(2\,\alpha^2 - 1) = \cos^{-1}(2\cos^2 y - 1)$$

Applying double angle identity $\cos 2y = 2\cos^2 y - 1$, we have

$$\cos^{-1}(2\cos^2 y - 1) = \cos^{-1}(\cos 2y) = 2y$$

Substituting back for y, we get RHS

$$2y = 2\cos^{-1}\alpha$$

Solution-17: Applying **Inv-6.6.13** $\tan^{-1}\alpha + \tan^{-1}\beta = \tan^{-1}\left(\frac{\alpha+\beta}{1-\alpha\beta}\right)$, we have

$$\tan^{-1}\frac{1}{2} + \tan^{-1}\frac{2}{5} = \tan^{-1}\left(\frac{\frac{1}{2}+\frac{2}{5}}{1-\frac{1}{2}\frac{2}{5}}\right)$$

$$= \tan^{-1}\left(\frac{\frac{5+4}{10}}{\frac{5-1}{5}}\right)$$

$$= \tan^{-1}\left(\frac{9}{10}\frac{5}{4}\right) = \tan^{-1}\left(\frac{9}{8}\right)$$

Solution-18: Applying **Inv-6.6.14** $\tan^{-1}\alpha - \tan^{-1}\beta = \tan^{-1}\frac{\alpha-\beta}{1+\alpha\beta}$, we have

$$\tan^{-1}\frac{1}{2} - \tan^{-1}\frac{2}{5} = \tan^{-1}\left(\frac{\frac{1}{2}-\frac{2}{5}}{1+\frac{1}{2}\frac{2}{5}}\right)$$

$$= \tan^{-1}\left(\frac{\frac{5-4}{10}}{\frac{5+1}{5}}\right)$$

$$= \tan^{-1}\left(\frac{1}{10}\frac{5}{6}\right) = \tan^{-1}\left(\frac{1}{12}\right)$$

Solution-19: Refer the identities **Inv-6.6.1, Inv-6.6.2, Inv-6.6.3-A, Inv-6.6.3-B, Inv-6.6.3-C, Inv-6.6.3-D** of this chapter.

Solution-20: Refer the identities **Inv-6.6.4, Inv-6.6.5, Inv-6.6.6** of this chapter.

Solution-21: Refer the identities **Inv-6.6.7, Inv-6.6.8, Inv-6.6.9** of this chapter.

Solution-22: Refer the identities **Inv-6.6.10, Inv-6.6.11, Inv-6.6.12** of this chapter.

Solution-23: Refer the identities **Inv-6.6.13, Inv-6.6.14** of this chapter.

Solution-24: Refer the identities **Inv-6.6.15, Inv-6.6.16, Inv-6.6.17, Inv-6.6.18** of this chapter.

7 Chapter 7: Sines/Cosines Rules, Heron's Formula

7.1 About Laws of Cosines/Sines

The law of Sines and Cosines allow us to solve for all kinds of triangles including oblique triangles (non-right triangles).

Law of Sines and Cosines are part of traditional trigonometry or Euclidean Geometry, which are not included in Calculus.

Law of Cosines is more popular than law of Sines, or law of Tangents.

7.2 The Law of Cosines

The law of Cosines is also known as Cosines Rule or Cosines Formula.

In the figure depicted below, a, b, and c are the lengths of the sides of a triangle; and α, β, and γ are respectively the opposite angles of these sides.

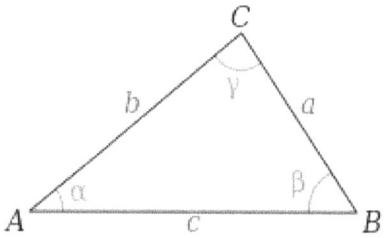

Cosines law: For any triangle including oblique triangle with given sides b and c and their included angle α, the Cosines law states:

$$a^2 = b^2 + c^2 - 2bc \cos \alpha$$

Similarly, Cosine law can be stated as:

$$b^2 = a^2 + c^2 - 2ac \cos \beta$$

$$c^2 = b^2 + a^2 - 2ba \cos \gamma$$

The Cosine law can also be applied to find any unknown angle, if all the three sides are known:

$$\cos \alpha = \frac{b^2 + c^2 - a^2}{2bc}$$

Remarks: Cosine rule relates the three sides of a triangle to Cosine of one of its angles. Cosines rule is used to:

- *find the unknown side of a triangle, if two sides and included angle are known.*
- *find any unknown angle of a triangle, if three sides are known.*

Example-1: Find the length x in following figure.

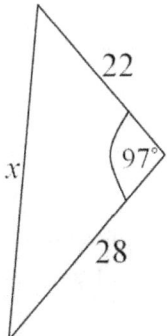

Solution: The triangle is not right-angled, we know two sides as well as included angle, and we need to find out unknown side, so we use the Cosine rule.

Cosine rule is $a^2 = b^2 + c^2 - 2bc \cos \alpha$

Filling in the appropriate values, we have

$$x^2 = 22^2 + 28^2 - 2\,(22)\,(28)\cos 97°$$

Note that, side b and c can be interchanged with each other without affecting the result.

We convert 97° to a smaller angle using complementary identity: $\cos(90°+x) = -\sin x$; and refer the appendix to find the value:

$\cos 97° = -\sin 7° = -0.122$ (refer the table in appendix for value of $\sin 7°$)

$$x^2 = 484 + 784 + 150.3 = 1418.3$$

Therefore, $x = 37.66$

Example-2: Find the angle P in following figure.

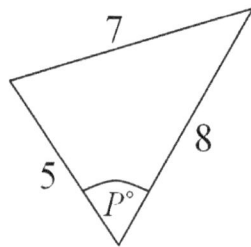

Solution: The given triangle is not right-angled, we know three sides, and we need to find included angle of known sides, so we use the Cosine Rule.

$$\cos P° = \frac{b^2 + a^2 - c^2}{2ba} = \frac{5^2 + 8^2 - 7^2}{2(5)(8)} = \frac{40}{80} = \frac{1}{2}$$

We know the $\cos 60° = \dfrac{1}{2}$, so angle $P = 60°$

Example-3: Find an unknown side b and an unknown angle C in following diagram.

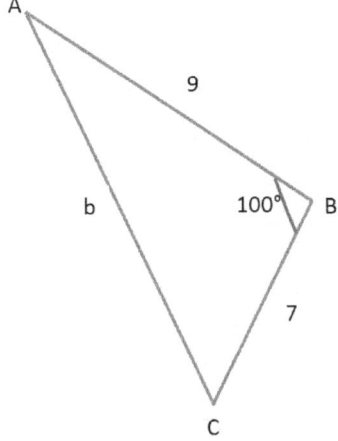

Solution: The triangle is not right-angled, we know two sides as well as included angle, and we need to find out unknown side, so we use the Cosine Rule.

Cosine rule is $b^2 = a^2 + c^2 - 2ac \cos B$

As given, a = 7, c = 9 and angle B = 100°, (sides a and c can be interchanged here)

So, $b^2 = 7^2 + 9^2 - 2 \times 7 \times 9 \times \cos 100°$

$= 49 + 81 - 126 \cos 100° = 130 - 126 \cos 100°$

Therefore, $b = \sqrt{(130 - 126 \cos 100°)}$

Convert the angle 100° to smaller angle and find the value in appendix. $\cos 100° = - \sin 10° = - 0.1736$ (refer the table in appendix for value of $\sin 10°$)

$b = \sqrt{(130 + 21.87)} = \sqrt{151.87} = 12.324$ (approx.)

After finding the side b, we can find unknown angle C as follows:

$$2ba \cos C = b^2 + a^2 - c^2$$

$$\cos C = \frac{b^2 + a^2 - c^2}{2ba} = \frac{151.87 + 49 - 81}{2(12.324)(7)} = \frac{119.87}{172.536} = 0.6947$$

Referring the table in appendix, we find that Cosine of 46° is 0.6947. So, angle C = 46° (approx.)

7.3   The Law of Sines

The law of Sines is also known as Sines Rule or Sines Formula.

In the figure depicted below, *a*, *b*, and *c* are the lengths of the sides of a triangle, and *A*, *B*, and *C* are the opposite angles respectively for these sides.

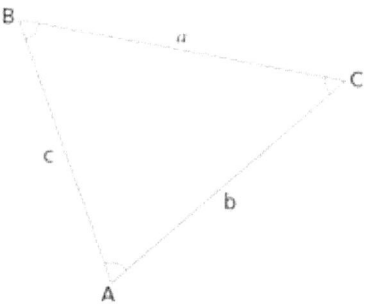

Sines law states that the sides of a triangle are to one another in the same ratio as the Sines of their opposite angles.

$$a : b : c = \sin A : \sin B : \sin C$$

This implies,

$$\frac{a}{\text{Sin A}} = \frac{b}{\text{Sin B}} = \frac{c}{\text{Sin C}}$$

It is also specified as reciprocals:

$$\frac{\text{Sin A}}{a} = \frac{\text{Sin B}}{b} = \frac{\text{Sin C}}{c}$$

Remarks: If sides of a triangle and one opposite angle between those two sides are known, then apply Sines rule to find other side or angle. Cosine rule uses included angle, whereas Sines rule uses opposite angle.

Example-1: Find the angle m in figure below.

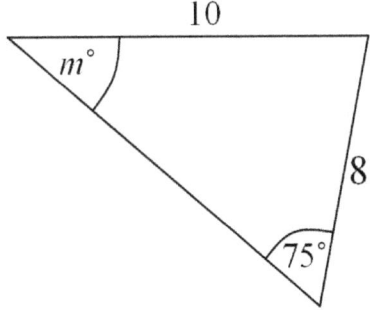

Solution: The known quantities are two sides and one opposite angle, so we apply Sines rule as follows:

$$\frac{\text{Sin A}}{a} = \frac{\text{Sin B}}{b}$$

Filling in the values, we have

$$\frac{\text{Sin m}}{8} = \frac{\text{Sin 75°}}{10}$$

$$\sin m = \frac{8\,(0.966)}{10} = 0.7728$$

Therefore, m = 51° (approx.)

Example-2: Find the side x below in figure.

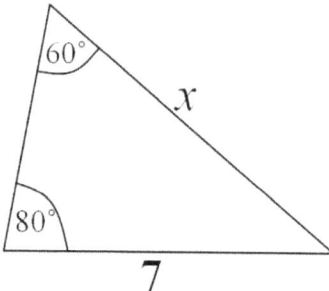

Solution: The known quantities are one side and two opposite angles, so we apply Sines rule as follows:

$$\frac{\text{Sin A}}{a} = \frac{\text{Sin B}}{b}$$

Filling in the values, we have

$$\frac{x}{\text{Sin 80°}} = \frac{7}{\text{Sin 60°}}$$

Therefore, $x = \dfrac{7\,(0.985)}{0.866} = 7.962$ (approx.)

7.4   The Law of Tangents

In the figure depicted below, a, b, and c are the lengths of the sides of a triangle, and α, β, and γ are respectively the opposite angles of these sides.

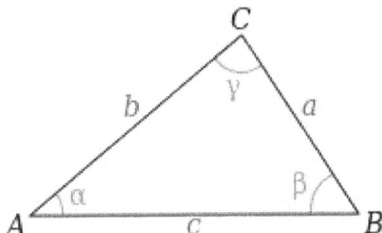

The law of Tangents is the relationship between the Tangents of two angles of a triangle and the lengths of the opposing sides.

With respect to give figure, the law of Tangents is expressed as:

$$\frac{a+b}{a-b} = \frac{\tan\left[\frac{1}{2}(\alpha + \beta)\right]}{\tan\left[\frac{1}{2}(\alpha - \beta)\right]}$$

Alternatively,

$$\frac{a-b}{a+b} = \frac{\tan\left[\frac{1}{2}(\alpha - \beta)\right]}{\tan\left[\frac{1}{2}(\alpha + \beta)\right]}$$

Remarks: This formula is not so much used as compared to the law of Sines or the law of Cosines.

Example-1: Find the side b in following figure.

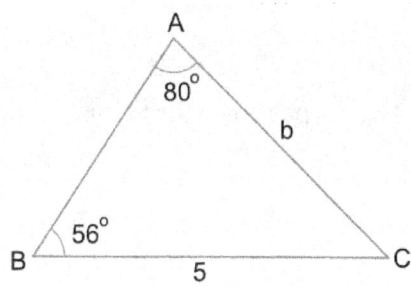

Solution: We use the Tangent rule $\dfrac{a+b}{a-b} = \dfrac{\tan\left[\frac{1}{2}(\alpha+\beta)\right]}{\tan\left[\frac{1}{2}(\alpha-\beta)\right]}$

As given, $a = 5$, A=80° and B =56°

So, $\dfrac{A+B}{2} = \dfrac{80°+56°}{2} = 68°$ and $\dfrac{A-B}{2} = \dfrac{80°-56°}{2} = 12°$

Filling in the values as per Tangent rule:

$$\dfrac{5+b}{5-b} = \dfrac{\tan 68°}{\tan 12°}$$

Cross multiplying the expressions, we have

$(5 + b)\tan 12° = (5 - b)\tan 68°$

$5\tan 12° + b\tan 12° = 5\tan 68° - b\tan 68°$

$b\tan 68° + b\tan 12° = 5\tan 68° - 5\tan 12°$

$b(\tan 68° + \tan 12°) = 5(\tan 68° - \tan 12°)$

Referring the Appendix for tan 68° and tan 12°, we get

$$b = \dfrac{5\,(2.475 - 0.213)}{(2.475 + 0.213)} = \dfrac{11.31}{2.69} = 4.21$$

7.5 Area of any Triangle

In this section, two types of rules for finding the area are considered: First rule for a right triangle and second rule for of an oblique triangle.

7.5.1 Area of a Right Triangle

If the base and perpendicular height of a triangle are known, then area of right triangle is calculated using the formula:

$$\text{Area} = \frac{1}{2} \times \text{base} \times \text{perpendicular}$$

So, area of a right triangle is half of product of its base and perpendicular.

This formula to find area is applicable only for right triangle, where perpendicular and base are known.

7.5.2 Area of an Oblique Triangle

In any triangle, right or oblique, area can be found by Sine trigonometric function.

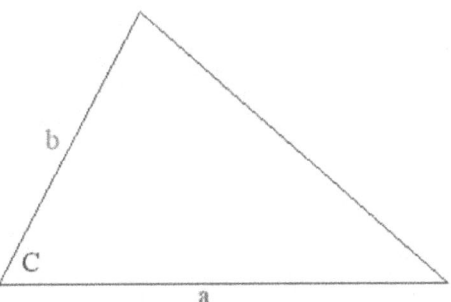

Area of any triangle with two sides a, b and included angle C in terms of trigonometric function is as follows:

$$\text{Area} = \frac{1}{2} \times a \times b \times \sin C$$

Remarks: This formula can also be used to find the area for a right triangle. In that case, angle C would be 90° and sin 90° =1 leads to formula of area for right triangle.

Therefore, for a right triangle, Area $= \frac{1}{2} \times a \times b$

Verification of formula: Area $= \frac{1}{2} \times a \times b \times \sin C$

To verify the area in terms of trigonometric function, let us draw a perpendicular on base from opposite vertex. Let the perpendicular be h as depicted in figure below.

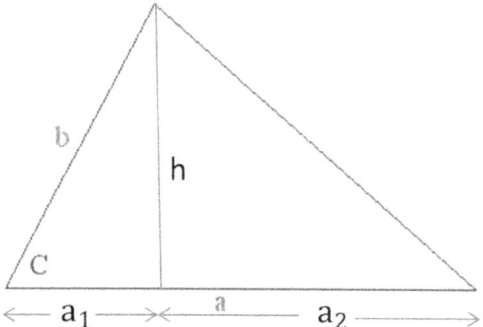

As we have two right triangles formed due to perpendicular h, the complete area of an oblique triangle can be specified in terms of sum of two right triangles.

Area $= \frac{1}{2} \times a_1 \times h + \frac{1}{2} \times a_2 \times h$

Area $= \frac{1}{2} \times (a_1 + a_2) \times h = \frac{1}{2} \times a \times h$

Using the trigonometric ratio, $\sin C = \frac{h}{b}$, we replace the perpendicular h with $(b \times \sin C)$

Area $= \frac{1}{2} \times a \times b \times \sin C$

7.5.3   Area using Heron's Formula

Heron's formula (also called Hero's formula) is applied to find the area of any triangle (right or oblique), if all the three sides are known.

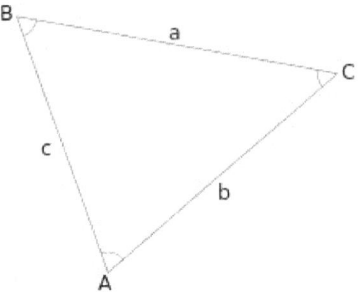

In any triangle, if the sides are a, b, c then Heron's formula is stated as:

$$\text{Area } (\Delta) = \sqrt{s\,(s-a)(s-b)(s-c)}$$

where s is the semi-perimeter of the triangle expressed as,

$$s = \frac{a+b+c}{2}$$

Example-1: Let ΔABC be the triangle with sides a=6 cm, b=12 cm and c=14 cm. Find its area.

Solution: Semi-perimeter $s = \dfrac{6+12+14}{2} = 16$ cm

$$\text{Area} = \sqrt{16\,(16-6)(16-12)(16-14)}$$

$$= \sqrt{16\,(10)(4)(2)} = \sqrt{1280} = 35.78 \text{ cm}^2$$

Worksheet-17

Exercise-1: Explain the Cosines law.

Exercise-2: Find the length x in following figure.

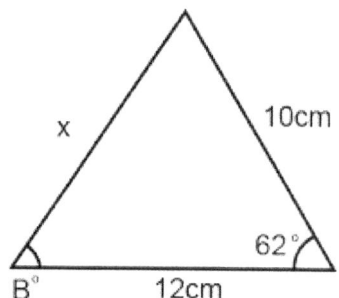

Exercise-3: Find the angle θ in following figure.

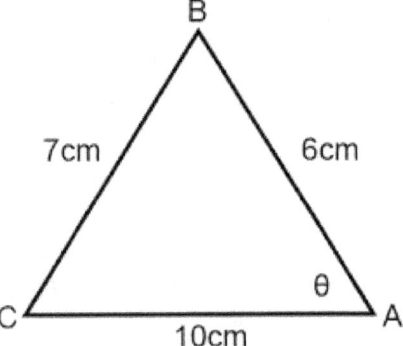

Exercise-4: Find an unknown side b and an unknown angle C in following diagram.

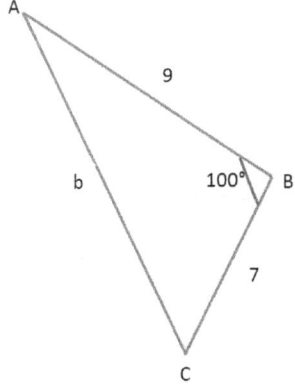

Exercise-5: Explain the Sines law.

Exercise-6: Find the angle m in figure below.

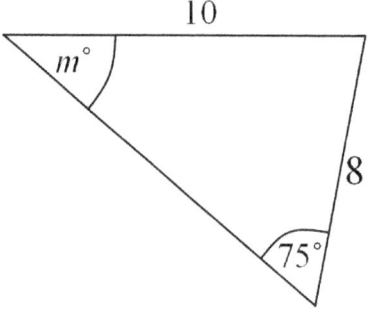

Exercise-7: Find the side *x* in figure below.

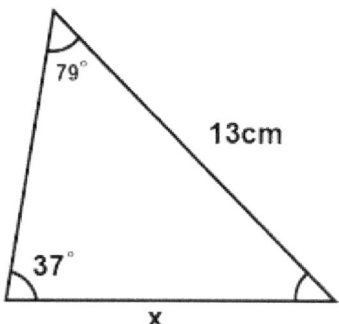

Exercise-8: Explain the law of Tangents.

Exercise-9: Find the side b in following figure.

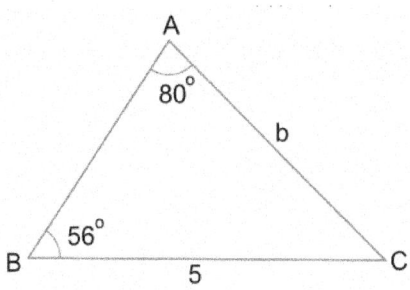

Exercise-10: Explain Heron's formula with respect to following figure.

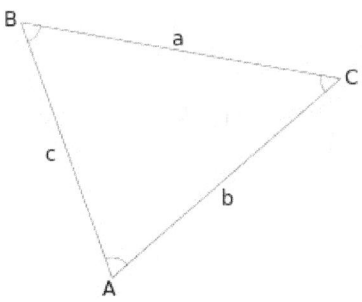

Exercise-11: Let $\triangle ABC$ be the triangle with sides a=12 cm, b=6 cm and c=14 cm. Find its area.

Solutions-Worksheet-17

Solution-1: For any triangle including oblique triangles, if the sides b and c and their included angle α are given, then the Cosines law states:

$$a^2 = b^2 + c^2 - 2bc \cos \alpha$$

Solution-2:

Cosine rules is $a^2 = b^2 + c^2 - 2bc \cos \alpha$

Filling in the appropriate values, we have

$$x^2 = 10^2 + 12^2 - 2\,(10)\,(12) \cos 62°$$

Note that, side b and c can be interchanged with each other without affecting the result.

Referring the value of $\cos 62° = 0.469$

$x^2 = 100 + 144 - 112.56 = 131.44$

Therefore, $x = 11.465$

Solution-3: The triangle is not right-angled, we know three sides, and we need to find included angle of known sides, so we use the Cosine Rule.

$$\cos \theta = \frac{b^2 + a^2 - c^2}{2ba} = \frac{6^2 + 10^2 - 7^2}{2(6)(10)} = \frac{87}{120} = 0.725$$

Referring the table for Cosine angle $\theta = 44°$ (approx.)

Solution-4: The triangle is not right-angled, we know two sides as well as included angle, and we need to find out unknown side, so we use the Cosine Rule.

Cosine rule is $b^2 = a^2 + c^2 - 2ac \cos B$

As given, a = 7, c = 9 and angle B = 100°, (sides a and c can be interchanged here)

So, $b^2 = 7^2 + 9^2 - 2 \times 7 \times 9 \times \cos 100°$

$$= 49 + 81 - 126 \cos 100° = 130 - 126 \cos 100°$$

Therefore, $b = \sqrt{(130 - 126 \cos 100°)}$

Convert the angle 100° to acute angle and find the value in appendix.
$\cos 100° = -\sin 10° = -0.1736$ (refer the appendix for value of sin 10°)

$$b = \sqrt{(130 + 21.87)} = \sqrt{151.87} = 12.324 \text{ (approx.)}$$

After finding the side b, we can find unknown angle γ as follows:

$2ba \cos C = b^2 + a^2 - c^2$

$$\cos C = \frac{b^2 + a^2 - c^2}{2ba} = \frac{151.87 + 49 - 81}{2(12.324)(7)} = \frac{119.87}{172.536} = 0.6947$$

Referring the appendix for angle of cos, we find that Cosine of 46° is the value 0.6947

So, side b = 12.324 and angle C = 46°

Solution-5: Sines law states that the sides of a triangle are to one another in the same ratio as the Sines of their opposite angles.

$a : b : c = \sin A : \sin B : \sin C$

This implies, $\dfrac{a}{\text{Sin A}} = \dfrac{b}{\text{Sin B}} = \dfrac{c}{\text{Sin C}}$

Solution-6: The known quantities are two sides and one opposite angle, so we apply Sines rule as follows:

$$\dfrac{\text{Sin A}}{a} = \dfrac{\text{Sin B}}{b}$$

Filling in the values, we have

$$\dfrac{\text{Sin m}}{8} = \dfrac{\text{Sin 75}°}{10}$$

$$\sin m = \dfrac{8\,(0.966)}{10} = 0.7728$$

Therefore, m = 51° (approx.)

Solution-7: The known quantities are one side and two opposite angle, so we apply Sines rule as follows:

$$\dfrac{\text{Sin A}}{a} = \dfrac{\text{Sin B}}{b}$$

Filling in the values, we have

$$\dfrac{x}{\text{Sin 79}°} = \dfrac{13}{\text{Sin 37}°}$$

Therefore, $x = \dfrac{13\,(0.9816)}{0.6018} = 21.20$ cm (approx.)

Solution-8: The law of Tangents is the relationship between the Tangents of two angles of a triangle and the lengths of the opposing sides.

The law of Tangents is expressed as: $\dfrac{a+b}{a-b} = \dfrac{\tan\,[\frac{1}{2}(\alpha+\beta)]}{\tan\,[\frac{1}{2}(\alpha-\beta)]}$

Solution-9: We use the Tangent rule $\dfrac{a+b}{a-b} = \dfrac{\tan\,[\frac{1}{2}(\alpha+\beta)]}{\tan\,[\frac{1}{2}(\alpha-\beta)]}$

As given, a = 5, A=80° and B =56°

So, $\dfrac{A+B}{2} = \dfrac{80°+56°}{2} = 68°$ and $\dfrac{A-B}{2} = \dfrac{80°-56°}{2} = 12°$

Filling in the values as per Tangent rule:

$$\frac{5+b}{5-b} = \frac{\tan 68°}{\tan 12°}$$

Cross multiplying the expressions, we have

$(5 + b)\tan 12° = (5 - b)\tan 68°$

$5\tan 12° + b\tan 12° = 5\tan 68° - b\tan 68°$

$b\tan 68° + b\tan 12° = 5\tan 68° - 5\tan 12°$

$b(\tan 68° + \tan 12°) = 5(\tan 68° - \tan 12°)$

$b = \dfrac{5(\tan 68° - \tan 12°)}{(\tan 68° + \tan 12°)}$

Referring the table in appendix for tan 68° and tan 12°, we get

$b = \dfrac{5(2.475 - 0.213)}{(2.475 + 0.213)} = \dfrac{11.31}{2.69} = 4.21$

Solution-10: Heron's formula is used to find the area of any triangle (right or oblique), if all the three sides are known.

In any triangle, if the sides are a, b, c then

$$\text{Area} = \sqrt{s\,(s-a)(s-b)(s-c)}$$

where s is the semi-perimeter of the triangle, $s = \dfrac{a+b+c}{2}$

Solution-11: Semi-perimeter $s = \dfrac{12+6+14}{2} = 16$ cm

$$\text{Area} = \sqrt{s\,(s-a)(s-b)(s-c)}$$

Filling in the values,

$$\text{Area} = \sqrt{16\,(16-12)(16-6)(16-14)}$$

$$= \sqrt{16\,(4)(10)(2)} = \sqrt{1280} = 35.78 \text{ cm}^2$$

Appendix 1- Table

Angle (deg)	$sin(\theta)$	$cos(\theta)$	$tan(\theta)$
0	0.0000	1.0000	0.0000
1	0.0175	0.9998	0.0175
2	0.0349	0.9994	0.0349
3	0.0523	0.9986	0.0524
4	0.0698	0.9976	0.0699
5	0.0872	0.9962	0.0875
6	0.1045	0.9945	0.1051
7	0.1219	0.9925	0.1228
8	0.1392	0.9903	0.1405
9	0.1564	0.9877	0.1584
10	0.1736	0.9848	0.1763
11	0.1908	0.9816	0.1944
12	0.2079	0.9781	0.2126
13	0.2250	0.9744	0.2309
14	0.2419	0.9703	0.2493
15	0.2588	0.9659	0.2679
16	0.2756	0.9613	0.2867
17	0.2924	0.9563	0.3057
18	0.3090	0.9511	0.3249
19	0.3256	0.9455	0.3443
20	0.3420	0.9397	0.3640
21	0.3584	0.9336	0.3839
22	0.3746	0.9272	0.4040
23	0.3907	0.9205	0.4245
24	0.4067	0.9135	0.4452
25	0.4226	0.9063	0.4663
26	0.4384	0.8988	0.4877
27	0.4540	0.8910	0.5095
28	0.4695	0.8829	0.5317
29	0.4848	0.8746	0.5543
30	0.5000	0.8660	0.5774
31	0.5150	0.8572	0.6009
32	0.5299	0.8480	0.6249
33	0.5446	0.8387	0.6494
34	0.5592	0.8290	0.6745
35	0.5736	0.8192	0.7002
36	0.5878	0.8090	0.7265
37	0.6018	0.7986	0.7536
38	0.6157	0.7880	0.7813
39	0.6293	0.7771	0.8098
40	0.6428	0.7660	0.8391
41	0.6561	0.7547	0.8693
42	0.6691	0.7431	0.9004
43	0.6820	0.7314	0.9325
44	0.6947	0.7193	0.9657
45	0.7071	0.7071	1.0000

Angle (deg)	sin(θ)	cos(θ)	tan(θ)
45	0.7071	0.7071	1.0000
46	0.7193	0.6947	1.0355
47	0.7314	0.6820	1.0724
48	0.7431	0.6691	1.1106
49	0.7547	0.6561	1.1504
50	0.7660	0.6428	1.1918
51	0.7771	0.6293	1.2349
52	0.7880	0.6157	1.2799
53	0.7986	0.6018	1.3270
54	0.8090	0.5878	1.3764
55	0.8192	0.5736	1.4281
56	0.8290	0.5592	1.4826
57	0.8387	0.5446	1.5399
58	0.8480	0.5299	1.6003
59	0.8572	0.5150	1.6643
60	0.8660	0.5000	1.7321
61	0.8746	0.4848	1.8040
62	0.8829	0.4695	1.8807
63	0.8910	0.4540	1.9626
64	0.8988	0.4384	2.0503
65	0.9063	0.4226	2.1445
66	0.9135	0.4067	2.2460
67	0.9205	0.3907	2.3559
68	0.9272	0.3746	2.4751
69	0.9336	0.3584	2.6051
70	0.9397	0.3420	2.7475
71	0.9455	0.3256	2.9042
72	0.9511	0.3090	3.0777
73	0.9563	0.2924	3.2709
74	0.9613	0.2756	3.4874
75	0.9659	0.2588	3.7321
76	0.9703	0.2419	4.0108
77	0.9744	0.2250	4.3315
78	0.9781	0.2079	4.7046
79	0.9816	0.1908	5.1446
80	0.9848	0.1736	5.6713
81	0.9877	0.1564	6.3138
82	0.9903	0.1392	7.1154
83	0.9925	0.1219	8.1443
84	0.9945	0.1045	9.5144
85	0.9962	0.0872	11.4301
86	0.9976	0.0698	14.3007
87	0.9986	0.0523	19.0811
88	0.9994	0.0349	28.6363
89	0.9998	0.0175	57.2900
90	1.0000	0.0000	undef

Appendix 2 – Trigonometric Identities

Negative identities:

$$\sin(-\theta) = -\sin\theta; \quad \operatorname{cosec}(-\theta) = -\operatorname{cosec}\theta;$$
$$\cos(-\theta) = \cos\theta; \quad \sec(-\theta) = \sec\theta;$$
$$\tan(-\theta) = -\tan\theta; \quad \cot(-\theta) = -\cot\theta;$$

Complementary Identities / Cofunction Identities:

$$\sin(90° - \theta) = \cos\theta$$
$$\cos(90° - \theta) = \sin\theta$$
$$\tan(90° - \theta) = \cot\theta$$
$$\cot(90° - \theta) = \tan\theta$$
$$\sec(90° - \theta) = \operatorname{cosec}\theta$$
$$\operatorname{cosec}(90° - \theta) = \sec\theta$$

Identities of 90°+θ:

$$\sin(90° + \theta) = \cos\theta$$
$$\cos(90° + \theta) = -\sin\theta$$
$$\tan(90° + \theta) = -\cot\theta$$
$$\cot(90° + \theta) = -\tan\theta$$
$$\sec(90° + \theta) = -\operatorname{cosec}\theta$$
$$\operatorname{cosec}(90° + \theta) = \sec\theta$$

Supplementary Identities of 180°−θ:

$\sin(180° − \theta) = \sin \theta$

$\cos(180° − \theta) = -\cos \theta$

$\tan(180° − \theta) = -\tan \theta$

$\cot(180° − \theta) = -\cot \theta$

$\sec(180° − \theta) = -\sec \theta$

$\csc(180° − \theta) = \csc \theta$

Identities of 180°+θ:

$\sin(180° + \theta) = -\sin \theta$

$\cos(180° + \theta) = -\cos \theta$

$\tan(180° + \theta) = \tan \theta$

$\cot(180° + \theta) = \cot \theta$

$\sec(180° + \theta) = -\sec \theta$

$\csc(180° + \theta) = -\csc \theta$

Identities of 360°−θ:

$\sin(360° − \theta) = -\sin \theta$

$\cos(360° − \theta) = \cos \theta$

$\tan(360° − \theta) = -\tan \theta$

$\cot(360° − \theta) = -\cot \theta$

$\sec(360° − \theta) = \sec \theta$

$\csc(360° − \theta) = -\csc \theta$

Identities of 360°+θ:

$\sin(360° + \theta) = \sin \theta$

$\cos(360° + \theta) = \cos \theta$

$$tan \ (360° + \theta) = tan \ \theta$$
$$cot \ (360° + \theta) = cot \ \theta$$
$$sec \ (360° + \theta) = sec \ \theta$$
$$cosec \ (360° + \theta) = cosec \ \theta$$

Identities of Sum/Difference of Angles:

$$\sin (\alpha + \beta) = \sin \alpha \cos \beta + \cos \alpha \sin \beta$$

$$\sin (\alpha - \beta) = \sin \alpha \cos \beta - \cos \alpha \sin \beta$$

$$\cos (\alpha + \beta) = \cos \alpha \cos \beta - \sin \alpha \sin \beta$$

$$\cos (\alpha - \beta) = \cos \alpha \cos \beta + \sin \alpha \sin \beta$$

$$\tan (\alpha + \beta) = \frac{\tan \alpha + \tan \beta}{1 - \tan \alpha \tan \beta} \ , \ \textit{where none of the angles } \alpha, \ \beta$$
$$\textit{and } (\alpha+\beta) \textit{ are an odd multiple of } \frac{\pi}{2}$$

$$\tan (\alpha - \beta) = \frac{\tan \alpha - \tan \beta}{1 + \tan \alpha \tan \beta} \ , \ \textit{where none of the angles } \alpha, \ \beta$$
$$\textit{and } (\alpha-\beta) \textit{ are an odd multiple of } \frac{\pi}{2}$$

$$\cot (\alpha + \beta) = \frac{\cot \alpha \cot \beta - 1}{\cot \alpha + \cot \beta} \ , \ \textit{where none of the angles } \alpha, \ \beta$$
$$\textit{and } (\alpha + \beta) \textit{ are a multiple of } \pi$$

$$\cot (\alpha - \beta) = \frac{\cot \alpha \cot \beta + 1}{\cot \beta - \cot \alpha} \ , \ \textit{where none of the angles } \alpha, \ \beta$$
$$\textit{and } (\alpha - \beta) \textit{ are a multiple of } \pi$$

Double Angle Identities:

$$\sin 2\alpha = 2 \sin \alpha \cos \alpha$$

$$\sin 2\alpha = \frac{2 \tan \alpha}{1 + \tan^2 \alpha}$$

$$\cos 2\alpha = \cos^2 \alpha - \sin^2 \alpha$$

$$\cos 2\alpha = 1 - 2 \sin^2 \alpha$$

$$\cos 2\alpha = 2 \cos^2 \alpha - 1$$

$$\cos 2\alpha = \frac{1 - \tan^2 \alpha}{1 + \tan^2 \alpha}$$

$$\tan 2\alpha = \frac{2 \tan \alpha}{1 - \tan^2 \alpha}$$

$$\cot 2\alpha = \frac{\cot^2 \alpha - 1}{2 \cot \alpha}$$

Half Angle Identities:

$$\sin \frac{\alpha}{2} = \pm \sqrt{\frac{1 - \cos \alpha}{2}}$$

$$\cos \frac{\alpha}{2} = \pm \sqrt{\frac{1 + \cos \alpha}{2}}$$

$$\tan \frac{\alpha}{2} = \frac{\sin \alpha}{1 + \cos \alpha}$$

Multiple Angle Formulas:

$$\sin 3\alpha = = 3 \sin \alpha \cos^2 \alpha - \sin^3 \alpha$$

$$\cos 3\alpha = \cos^3 \alpha - 3 \sin^2 \alpha \cos \alpha$$

$$\tan 3\alpha = \frac{3 \tan \alpha - \tan^3 \alpha}{1 - 3 \tan^2 \alpha}$$

$$\sin 4\alpha = 4 \sin \alpha \cos^3 \alpha - 4 \cos \alpha \sin^3 \alpha$$

Sum as Product Identities:

$$\sin(\alpha + \beta) + \sin(\alpha - \beta) = 2 \sin \alpha \cos \beta$$

$$\sin(\alpha + \beta) - \sin(\alpha - \beta) = 2 \cos \alpha \sin \beta$$

$$\cos(\alpha + \beta) + \cos(\alpha - \beta) = 2 \cos \alpha \cos \beta$$

$$\cos(\alpha + \beta) - \cos(\alpha - \beta) = -2 \sin \alpha \sin \beta$$

$$\sin a + \sin b = 2 \sin \frac{a+b}{2} \cos \frac{a-b}{2}$$

$$\sin a - \sin b = 2 \cos \frac{a+b}{2} \sin \frac{a-b}{2}$$

$$\cos a + \cos b = 2 \cos \frac{a+b}{2} \cos \frac{a-b}{2}$$

$$\cos a - \cos b = -2 \sin \frac{a+b}{2} \sin \frac{a-b}{2}$$

Product as Sum Identities:

$$2 \sin \alpha \cos \beta = \sin(\alpha + \beta) + \sin(\alpha - \beta)$$

$$2 \cos \alpha \sin \beta = \sin(\alpha + \beta) - \sin(\alpha - \beta)$$

$$2 \cos \alpha \cos \beta = \cos(\alpha + \beta) + \cos(\alpha - \beta)$$

$$-2 \sin \alpha \sin \beta = \cos(\alpha + \beta) - \cos(\alpha - \beta)$$

$$2 \sin \frac{a+b}{2} \cos \frac{a-b}{2} = \sin a + \sin b$$

$$2 \cos \frac{a+b}{2} \sin \frac{a-b}{2} = \sin a - \sin b$$

$$2 \cos \frac{a+b}{2} \cos \frac{a-b}{2} = \cos a + \cos b$$

$$-2 \sin \frac{a+b}{2} \sin \frac{a-b}{2} = \cos a - \cos b$$

10 Appendix 3- General Solutions

General Solutions:

$\sin \alpha = \sin \beta$ implies $\alpha = n\pi + (-1)^n \beta$

$\cos \alpha = \cos \beta$ implies $\alpha = 2n\pi \pm \beta$

$\tan \alpha = \tan \beta$ implies $\alpha = n\pi + \beta$ (where $\alpha, \beta \neq (2n+1)\dfrac{\pi}{2}$)

$\sin^2 \alpha = \sin^2 \beta$ implies $\alpha = n\pi \pm \beta$

$\cos^2 \alpha = \cos^2 \beta$ implies $\alpha = n\pi \pm \beta$

$\tan^2 \alpha = \tan^2 \beta$ implies $\alpha = n\pi \pm \beta$ (where $\alpha, \beta \neq (2n+1)\dfrac{\pi}{2}$)

[where $n \in Z$ (integer set) and β is first principal solution]

General Solutions for specific values:

Case-1: $\sin \alpha = 0 \Rightarrow \alpha = n\pi$

Case-2: $\sin \alpha = 1 \Rightarrow \alpha = (4n+1)\dfrac{\pi}{2}$

Case-3: $\sin \alpha = -1 \Rightarrow \alpha = (4n-1)\dfrac{\pi}{2}$

Case-4: $\cos \alpha = 0 \Rightarrow \alpha = (2n+1)\dfrac{\pi}{2}$

Case-5: $\cos \alpha = 1 \Rightarrow \alpha = 2n\pi$

Case-6: $\cos \alpha = -1 \Rightarrow \alpha = (2n+1)\pi$

Case-7: $\tan \alpha = 0 \Rightarrow \alpha = n\pi$

(where $n \in Z$, set of integers)

Appendix 4- Inverse Trigonometric Functions

Range/ Principal Branch of Inverse Trigonometric Functions

Function name	Function	Domain	Pr. Branch
inverse Sine	$\sin^{-1} x$	$[-1, 1]$	$[-\frac{\pi}{2}, \frac{\pi}{2}]$
inverse Cosine	$\cos^{-1} x$	$[-1, 1]$	$[0, \pi]$
inverse Tangent	$\tan^{-1} x$	$(-\infty, \infty)$	$(-\frac{\pi}{2}, \frac{\pi}{2})$
inverse Cotangent	$\cot^{-1} x$	$(-\infty, \infty)$	$(0, \pi)$
inverse Secant	$\sec^{-1} x$	$(-\infty, -1]$ and $[1, \infty)$	$[0, \frac{\pi}{2})$ and $(\frac{\pi}{2}, \pi]$
inverse Cosecant	$\operatorname{cosec}^{-1} x$	$(-\infty, -1]$ and $[1, \infty)$	$[-\frac{\pi}{2}, 0)$ and $(0, \frac{\pi}{2}]$

Forward-Inverse Identities

$$\sin(\sin^{-1}(x)) = x$$
$$\cos(\sin^{-1}(x)) = \sqrt{1 - x^2}$$
$$\tan(\sin^{-1}(x)) = \frac{x}{\sqrt{1 - x^2}}$$
$$\sin(\cos^{-1}(x)) = \sqrt{1 - x^2}$$
$$\cos(\cos^{-1}(\beta)) = x$$

$$\tan(\cos^{-1}(x)) = \frac{\sqrt{1-x^2}}{x}$$

$$\sin(\tan^{-1}(x)) = \frac{x}{\sqrt{1+x^2}}$$

$$\cos(\tan^{-1}(x)) = \frac{1}{\sqrt{1+x^2}}$$

$$\tan(\tan^{-1}(x)) = x$$

Inverse-Forward Identities

$$\sin^{-1}(\cos x) = \frac{\pi}{2} - x, \text{ for } x \in [0, \pi]$$

$$\cos^{-1}(\sin x) = \frac{\pi}{2} - x, \text{ for } x \in \left[-\frac{\pi}{2}, \frac{\pi}{2}\right]$$

$$\tan^{-1}(\cot x) = \frac{\pi}{2} - x, \text{ for } x \in (0, \pi)$$

$$\cot^{-1}(\tan x) = \frac{\pi}{2} - x, \text{ for } x \in \left(-\frac{\pi}{2}, \frac{\pi}{2}\right)$$

$$\sec^{-1}(\operatorname{cosec} x) = \frac{\pi}{2} - x, \text{ for } x \in \left(0, \frac{\pi}{2}\right] \text{ and } \left[-\frac{\pi}{2}, 0\right)$$

$$\operatorname{cosec}^{-1}(\sec x) = \frac{\pi}{2} - x, \text{ for } x \in \left[0, \frac{\pi}{2}\right) \text{ and } \left(\frac{\pi}{2}, \pi\right]$$

Reciprocal Identities of Inverse

$$\sin^{-1}\frac{1}{\beta} = \operatorname{cosec}^{-1}\beta, \text{ where } \beta \geq 1 \text{ and } \beta \leq -1$$

$$\operatorname{cosec}^{-1}\frac{1}{\beta} = \sin^{-1}\beta, \text{ where } \beta \geq -1 \text{ and } \beta \leq 1$$

$$\cos^{-1}\frac{1}{\beta} = \sec^{-1}\beta, \text{ where } \beta \geq 1 \text{ and } \beta \leq -1$$

$$\sec^{-1}\frac{1}{\beta} = \cos^{-1}\beta, \text{ where } \beta \geq -1 \text{ and } \beta \leq 1$$

$$\tan^{-1}\frac{1}{\beta} = \frac{\pi}{2} - \tan^{-1}\beta = \cot^{-1}\beta \text{ , where } \beta > 0$$

$$\tan^{-1}\frac{1}{\beta} = \frac{-\pi}{2} - \tan^{-1}\beta = \cot^{-1}\beta - \pi \text{ ,where } \beta < 0$$

$$\cot^{-1}\frac{1}{\beta} = \frac{\pi}{2} - \cot^{-1}\beta = \tan^{-1}\beta \text{ , for } \beta > 0$$

$$\cot^{-1}\frac{1}{\beta} = \frac{3\pi}{2} - \cot^{-1}\beta = \tan^{-1}\beta + \pi \text{ , for } \beta < 0$$

Negative Identities of Inverse

$$\sin^{-1}(-\beta) = -\sin^{-1}\beta \text{ ; where } \beta \in [-1, 1]$$
$$\csc^{-1}(-\beta) = -\csc^{-1}\beta \text{ ; where } |\beta| \geq 1$$
$$\tan^{-1}(-\beta) = -\tan^{-1}\beta \text{ ; where } \beta \in \mathbf{R}$$

Negative Supplementary Identities of Inverse

$$\cos^{-1}(-\beta) = \pi - \cos^{-1}\beta \text{ ; where } \beta \in [-1, 1]$$
$$\sec^{-1}(-\beta) = \pi - \sec^{-1}\beta \text{ ; where } |\beta| \geq 1$$
$$\cot^{-1}(-\beta) = \pi - \cot^{-1}\beta \text{ ; where } \beta \in \mathbf{R}$$

Complementary Identities of Inverse

$$\sin^{-1}\beta + \cos^{-1}\beta = \frac{\pi}{2} \text{ ; where } \beta \in [-1, 1]$$
$$\tan^{-1}\beta + \cot^{-1}\beta = \frac{\pi}{2} \text{ ; where } \beta \in \mathbf{R}$$
$$\csc^{-1}\beta + \sec^{-1}\beta = \frac{\pi}{2} \text{ ; where } |\beta| \geq 1$$

Sum/Difference Identities of Inverse

$$\tan^{-1}\alpha \pm \tan^{-1}\beta = \tan^{-1}\left(\frac{\alpha \pm \beta}{1 \mp \alpha\beta}\right)$$

$$\sin^{-1}\alpha \pm \sin^{-1}\beta = \sin^{-1}(\alpha\sqrt{(1-\beta^2)} \pm \beta\sqrt{(1-\alpha^2)})$$

$$\cos^{-1}\alpha \pm \cos^{-1}\beta = \cos^{-1}(\alpha\beta \mp \sqrt{(1-\alpha^2)(1-\beta^2)})$$

$$\cot^{-1}\alpha \pm \cot^{-1}\beta = \cot^{-1}\left(\frac{\alpha\beta \mp 1}{\beta \pm \alpha}\right)$$

Double Angle Identities of Inverse

$$2\tan^{-1}\alpha = \sin^{-1}\frac{2\alpha}{1+\alpha^2}, \text{ where } |\alpha| \leq 1$$

$$2\tan^{-1}\alpha = \cos^{-1}\frac{1-\alpha^2}{1+\alpha^2}, \text{ where } \alpha \geq 1$$

$$2\tan^{-1}\alpha = \tan^{-1}\frac{2\alpha}{1-\alpha^2}, \text{ where } -1 < \alpha < 1$$

$$\sin^{-1}(2\alpha\sqrt{1-\alpha^2}) = 2\sin^{-1}\alpha\,;\ \text{where } \frac{-1}{\sqrt{2}} \leq \alpha \leq \frac{1}{\sqrt{2}}$$

Half Angle Identities of Inverse

$$\sin^{-1}\alpha = 2\tan^{-1}\left(\frac{\alpha}{1+\sqrt{(1-\alpha^2)}}\right)$$

$$\cos^{-1}\alpha = 2\tan^{-1}\left(\frac{\sqrt{(1-\alpha^2)}}{1+\alpha}\right), \text{ if } -1 < \alpha \leq 1$$

$$\tan^{-1}\alpha = 2\tan^{-1}\left(\frac{\alpha}{1+\sqrt{(1+\alpha^2)}}\right)$$

12 Appendix 5- Sines/Cosines Rules

Sines/Cosines/Heron's Rules

Cosines law: $a^2 = b^2 + c^2 - 2bc \cos \alpha$

Sines Law: $a : b : c = \sin A : \sin B : \sin C$

Tangents Law: $\dfrac{a+b}{a-b} = \dfrac{\tan\left[\frac{1}{2}(\alpha + \beta)\right]}{\tan\left[\frac{1}{2}(\alpha - \beta)\right]}$

Heron's Formula: Area $(\Delta) = \sqrt{s\,(s-a)(s-b)(s-c)}$

where s is the semi-perimeter of Δ; $s = \dfrac{a+b+c}{2}$

13 Index

Kalisey Series

Delivering the highest quality academic materials

All essential concepts of trigonometry for the middle and high school students are covered herein.

Guaranteed learning of trigonometry for the students of middle & high schools

- **Saurya Singh**
Kalisey Academy Publication

For any further enquiries about the book and content, contact:
info@kalisey-softek.com